HALLEY'S
COMET
IN HISTORY

Dr. Halley.

HALLEY'S COMET IN HISTORY

by

Hermann Hunger

F. Richard Stephenson

Christopher B. F. Walker

Kevin K. C. Yau

edited by

F. R. Stephenson

and C. B. F. Walker

Published for the
Trustees of the British Museum by
British Museum Publications Limited

Published by British Museum Publications Ltd
46 Bloomsbury Street, London WC1B 3QQ

British Library Cataloguing in Publication Data
Halley's comet in history.
1. Halley's comet
I. Hunger, Hermann II. Stephenson, F. R.
III. Walker, C. B. F. IV. British Museum
523.6'4 QB723.H2

ISBN 0-7141-1118-X

Designed by Harry Green
Set in Monophoto Photina and printed by
Jolly & Barber Ltd, Rugby, Warwickshire

Front cover Halley's Comet watched by the English court in AD 1066, from the Bayeux Tapestry. Reproduced by special permission of the town of Bayeux

Half-title inscription
'At the command of my Lord and Lady may it go well.' This prayer to the gods was written on the edge of many Babylonian astronomical texts in the last four centuries BC

Frontispiece
Edmond Halley (1656–1742), by R. Phillips. Reproduced by permission of the Trustees of the National Portrait Gallery

Contents

ABBREVIATIONS

BMC + number	*British Museum Catalogue of Greek Coins*
LBAT	T. G. Pinches et al, *Late Babylonian Astronomical and Related Texts* (Providence 1955)
LT	Local time
SE	Seleucid Era
WA + number	Western Asiatic Antiquities + serial number of objects in the departmental collections

GLOSSARY

Aphelion	That point in a comet's orbit at which it is farthest from the sun
Asterism	A cluster of stars
Coma	The nebulous envelope around the head of a comet
Conjunction	An apparent proximity of two heavenly bodies
Ecliptic	The great circle of the celestial sphere which is the apparent orbit of the sun
Intercalary	(Of a month) inserted in the ordinary (twelve-month) year to harmonise it with the solar year
Occultation	The concealment of one heavenly body by another passing between it and the observer
Perihelion	That point in a comet's orbit at which it is nearest to the sun
Babylonian measures	1 cubit = 2 – 2.5°; 24 fingers = 1 cubit. 1 *kur* = 5 *pan*, 1 *pan* = 6 *sât*, 1 *sūt* = 6 *qa*, 1 *qa* = 0.84 litre

Preface

This book has been written to accompany an exhibition 'Halley's Comet in History' at the British Museum which illustrates the history of scientific observation of the comet up to the time of Halley himself. It lays particular emphasis on the Babylonian and Chinese records which are from the scientific point of view superior to any European observations before the fifteenth century AD. It was inspired by the discovery in the Museum's collections of Babylonian observations of the comet in 164 and 87 BC. These have proved to be of considerable interest for the study of the comet's orbit in antiquity.

The exhibition has been organised with the co-operation of the British Library. It will include a computerised explanation of the Babylonian cuneiform texts and a graphic display of astronomical observations. Computer equipment for the exhibition has been generously supplied by Personal Computers Ltd., Macintosh Centre, 218 Bishopsgate, London EC2.

The map of Babylonia and other illustrations have been drawn by Ann Searight.

We dedicate this book to the memory of Abraham Sachs whose pioneering efforts over more than thirty years have been fundamental to all study of the Babylonian astronomical diaries.

F. R. Stephenson
C. B. F. Walker
June 1985

CHAPTER 1

Introduction

Characteristics of bright comets

Bright comets appear fairly frequently, perhaps every five or ten years on the average, and readers may recall comets Ikeya-Seki (1965), Bennett (1970) and West (1976). However, in modern times with abundant artificial light illuminating the night sky these objects are not noticed to the same extent as they were even fifty years ago. A typical comet consists of a central nucleus containing frozen gases and rocky material which is surrounded by a diffuse head (or *coma*). Pointing away from the head is a lengthy luminous tail. To the unaided eye the tiny nucleus – if visible at all – resembles a star. Measurements made by astronomers indicate a size of only a few miles across. A sphere of ice even ten miles in diameter (larger than a typical cometary nucleus) would weigh less than a thousand-millionth of the mass of the Earth and this must be regarded as a likely upper limit. The coma has a misty appearance; its size may be colossal, perhaps more than one hundred thousand miles across. It is not unusual for the tail to extend for many millions of miles. Nevertheless practically all of the material lies in the nucleus; the coma and tail are virtually empty space.

Most comets spend much of their time at vast distances from the sun. In the cold depths of space a comet has neither coma nor tail. On account of its small size it would reflect so little sunlight that it would be invisible even in a large telescope. As the comet approaches the sun the solar radiation begins to melt and eventually vaporise the surface ices. Since the comet's gravity is so weak, the gases formed rapidly spread out into space forming a huge envelope known as the coma. Due to the vast size of the coma it reflects much sunlight and may become very brilliant indeed. The tail is formed from dust particles which were impurities in the ices.

Halley's Comet has its *perihelion* inside the orbit of the planet Venus. However, at *aphelion* (its furthest point from the sun) it is beyond the orbit of Neptune. It thus crosses the paths of seven planets including the Earth. At perihelion it moves at more than 120,000 mph or thirty-four miles every second. On the other hand near aphelion its speed is around 2000 mph, far less than that of an artificial satellite orbiting the Earth. Due to its immense speed as it swings around the sun, near perihelion the comet covers an angle of 180° in just over three months. It takes about seventy-five years to describe the remaining 180°.

Halley's Comet – or P/Halley as it is often denoted to indicate its periodic nature – is unique among bright comets in making periodic returns to the inner solar system. All other bright comets orbit the sun in many thousands or even millions of years. In some cases it is doubtful whether such visitors from outer space, once having left the sun's vicinity, will ever return again. The regularity of Halley's Comet is tantalising; the interval between two successive returns (seventy-five to eighty years) is quite close to the average human life-span, hence most people have but one brief opportunity to view the comet; it is rare indeed to have two. Although a bright comet which usually provides a worthwhile display, Halley's Comet is by no means the brightest known. It is often far outshone by its cousins from the depths of space, as happened at its last return in 1910.

Early speculation on comets

In ancient Greece, and to a lesser extent Rome, comets aroused widespread attention and led to much philosophical speculation. Around 500 BC we find the first Greek discussion of the nature of comets by Anaxagoras and Democritus. The former is noted for his belief that the sun was a red-hot stone 'larger than the Peloponnesus' – the peninsula to the west of Athens – which at least recognised that the sun was of substantial size. Democritus is well known for his 'atomic theory' of matter. These two philosophers held the view that comets were produced when planets or stars met or came into conjunction with one another, although some of their contemporaries considered comets to be planets.

Around 350 BC Aristotle gave powerful arguments against both ideas. In his treatise on *Meteorologica* he remarks, 'We ourselves have observed the planet Jupiter in conjunction with one of the stars in the Twins [Gemini] and hiding it completely, but no comet resulted.' Again he notes that the planets are always within the circle of the zodiac whereas many comets have been seen outside that circle. Despite these arguments, Aristotle believed that comets, like meteors, were produced by friction in the upper air caused by the daily revolution of the celestial sphere. The air thus became heated and burst into flames. If the fire travelled quickly a meteor was produced but a slow consuming fire was responsible for a comet. As proof that comets were fiery he remarked, 'When comets appear frequently and in considerable number the years are notoriously dry and windy', and cited examples from history.

Comets are not mentioned in the purely astronomical literature of early Europe. Ptolemy, one of the greatest astronomers of antiquity, who lived at Alexandria during the second century AD, did not investigate them. His great treatise the *Almagest*, which is still studied in detail today, is largely devoted to the understanding of the motion of the moon and planets. Perhaps he had recognised the random nature of the appearance of comets. However, it could be merely that Ptolemy, following Aristotle, believed comets to be no more than emanations from the Earth's atmosphere.

The astrological aspect was discussed in the first century AD by the Roman writer Pliny: 'If it resembles a pair of flutes it is a portent for the art of music, in the private parts of the constellations it portends immorality, if it forms an equilateral triangle or a rectangular quadrilateral in relation to certain positions of the fixed stars, it portends men of genius and a revival of learning, in the head of the Northern or the Southern Serpent it brings poisonings.'

His contemporary Seneca wrote extensively on comets, summarising earlier opinions, and reached more correct conclusions than Aristotle. In his opinion the fact that comets were not restricted to the zodiac circle did not disprove their planetary nature. He wrote, 'I do not think that a comet is a sudden fire but that it is among the eternal works of nature.' However, such was Aristotle's authority that his idea of comets as upper-air phenomena was generally accepted for nearly two thousand years. Only in the sixteenth century did Tycho Brahe, having made accurate observations of the comet of 1577, demonstrate that a comet moved in an orbit far beyond the moon and thus truly in the celestial vault.

Nothing is known regarding Egyptian or Babylonian speculation on the nature of comets. There are no known accounts of comets from ancient Egypt, but in fact few astronomical records of any kind survive from before the Roman period in Egypt. It would appear that despite the favourable remarks that Aristotle and other Greek writers made about the astronomers of Egypt only the most basic celestial observations were made, notably the rising of the bright star Sothis (Sirius). In contrast an extensive series of Babylonian observations have survived and these will be dealt with in chapter 2.

The Chinese made frequent sightings of comets and other astronomical phenomena from at least the eighth century BC. However, speculation on the nature of these objects seems to have been very restricted. In some cases they were ascribed to an imbalance of the Yin and the Yang. On a more sophisticated level they had been thought to have originated from the various planets. This was for example the belief of Chin Fang around 503 BC. The Chinese are credited with the discovery that a comet's tail tends to point directly away from the sun. This fact was recognised at least as early as AD 635, nearly a thousand years before it seems to have been noted in Europe.

Chinese interest in comets was mainly astrological. The second-century AD writer Wen Yin wrote the following commentary in the *Han-shu*:

> 'The three types of stars are *po* (bushy star), *hui* (broom star) and *ch'ang* (long star). Their astrological significance is similar but with slightly different size and shape. The rays of bushy stars are short; their light rays bush out in all directions. The light rays of broom stars are long and tufted like a broom. Long stars have one straight light ray, sometimes it stretches across the sky, sometimes it is 10 *chang* [about 100°], sometimes it is 3 *chang* [about 30°] and sometimes it is 2 *chang* [about 20°], without any regularity. According to the *Ta-fa*, bushy stars and broom stars are mostly for sweeping away with the old and spreading the new, or for fire disasters. Long stars mostly represent warfare.'

The early sightings of Halley's Comet

Of the ancient astronomers only the Babylonians and Chinese kept regular records of their observations over an extensive period of time; the motive was mainly astrological. Although Babylonian astronomical records are known to go back to about 700 BC, few records of comets have survived. There are only three records of Halley's Comet, two for 164 BC and the other for 87 BC (as recently discovered by Stephenson, Yau and Hunger). The Chinese have an almost complete record of Halley's Comet from 240 BC onwards. Up to and including the return of AD 1378 Far Eastern observations of Halley's Comet are much more detailed and accurate than those from the West. Accurate European positional measurements of Halley's Comet did not begin until the AD 1456 return, with the work of the Italian astronomer Paolo Toscanelli. On that occasion he made careful positional measurements of the comet – to a fraction of a degree – over a period of a month. The contemporaneous records from China and its cultural satellites Japan and Korea are of relatively low accuracy. At all later times precise measurements from the West have far outweighed the observations of Halley's Comet made in the traditional style by oriental astronomers. Nevertheless for any long-term investigation, the oriental data are indispensable.

The early positional data are particularly valuable in studying the long-term motion of the comet. During the past two millennia Halley's Comet has made several very close approaches to the Earth, notably in AD 837 when it passed by only about three million miles away (not much more than ten times the moon's distance). Such encounters severely perturb the motion of the comet. It is thus impossible to work out the precise orbital parameters for each return of Halley's

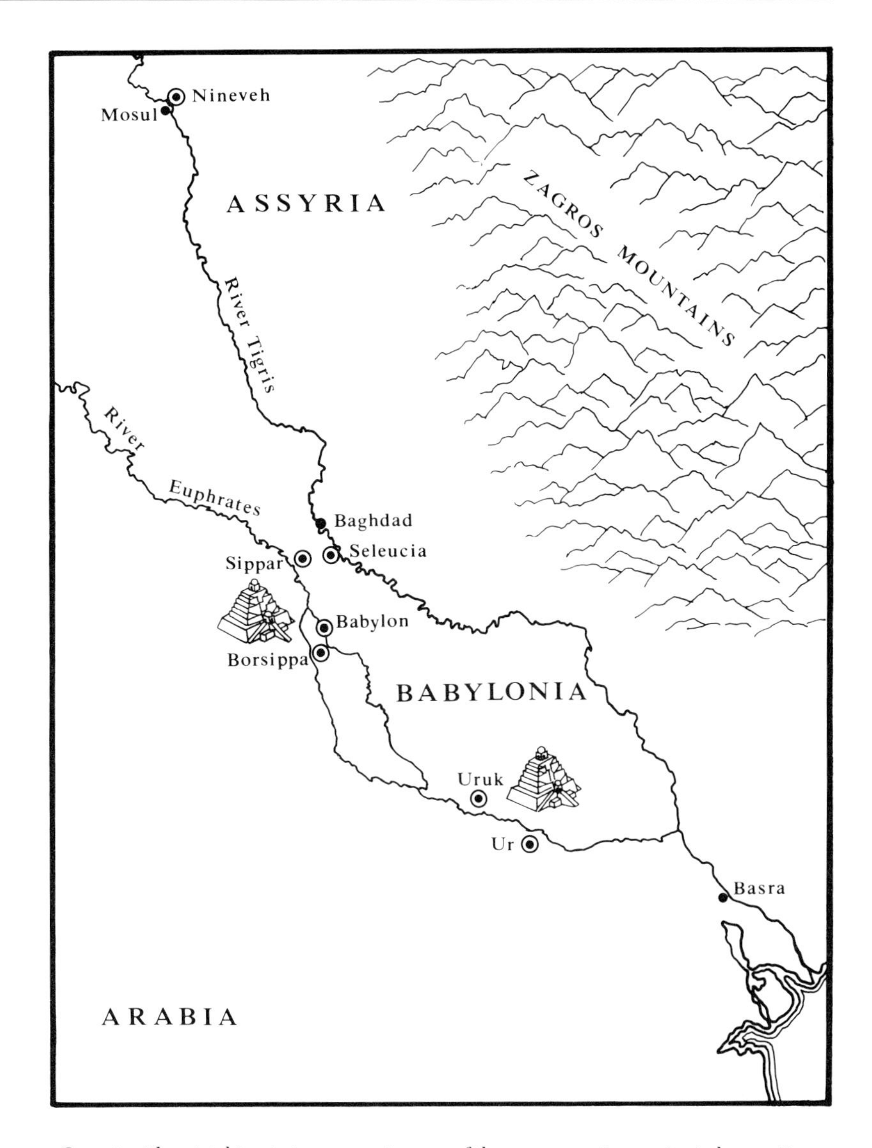

Map of Babylonia

Comet without taking into account some of the more precise ancient observations. The English astronomer John Hind in 1846 was the first to incorporate the Chinese observations in his solution and he extended his investigation back to 12 BC. This work was followed by Philip Cowell and Andrew Crommelin in their calculation of the 1910 return, and recently by Tao Kiang of the Dunsink Observatory, Dublin, and Donald Yeomans of the Jet Propulsion Laboratory, USA.

CHAPTER 2

Babylonian observations of Halley's Comet

The recovery of Babylonian astronomical texts

In 1705 Edmond Halley began his pamphlet, *A Synopsis of the Astronomy of Comets*, with a reference to the Babylonian (Chaldean) astronomers: 'The ancient Egyptians and Chaldeans (if we may credit Diodorus Siculus) by a long Course of Observations were able to predict the Apparitions of Comets. But since they are also said, by the Help of the same Arts, to have prognosticated Earthquakes and Tempests, 'tis past all Doubt, that their Knowledge in these Matters, was the Result rather of meer Astrological Calculation, than of any Astronomical Theories of the celestial Motions. And the Greeks who were the Conquerors of both those People, scarce found any other sort of Learning amongst them, than this. So that 'tis to the Greeks themselves as the Inventors (and especially to the Great Hipparchus) that we owe this Astronomy, which is now improv'd to such a height.' (Diodorus *Bibliotheca* I 81)

As lately as 1955 we would have been able to say no more than Halley himself about the Babylonian observation of comets. Nothing at all was known from the available Babylonian cuneiform texts, of which very few had been published. But in 1955 there appeared the first major publication of the late Babylonian astronomical archives. Copies of some 1600 Babylonian astronomical texts had been drawn in the 1890s by Theophilus G. Pinches of the British Museum and Father Johann N. Strassmaier, S.J., a Bavarian Assyriologist; but their work lay hidden in the archives of the British Museum for sixty years. In the early 1950s this material was entrusted to the late Professor Abraham Sachs of Brown University, USA, and was published as *Late Babylonian Astronomical and Related Texts*. The availability of this new source material has slowly transformed our knowledge of Babylonian astronomy. Until his death in 1983 Professor Sachs devoted his whole energies to the preparation of translations and explanations of these texts, and his unpublished papers have now been entrusted to Professor Hermann Hunger of Vienna University for eventual publication.

Some 2000 Late Babylonian astronomical texts are now known, mostly from the period 450 BC to AD 75, and it seems likely that the large majority of them come from a single archive at Babylon. A small group were found in the southern city of Uruk, and a few may have come from Borsippa and Sippar. The large archive from Babylon, now mostly in the British Museum, was probably found in unofficial excavations by the inhabitants of the nearby town of Hillah in the mid-1870s, and was purchased by the Museum through the dealer Spartali & Co. A small number were also found in the course of the Museum's own excavations at Babylon under the direction of Hormuzd Rassam in the early 1880s. Sadly no detailed record of these excavations was kept by Rassam, so that we remain ignorant about the precise spot in Babylon where the tablets were found. We do not know whether they formed part of a specialist library in a separate institution, whether they were housed in or near one of the major temples, or even whether they were in fact all found in a single place. It is commonly assumed that the Babylonian astronomers used the ziggurrat, the great stepped temple tower of Marduk at Babylon, as an observatory (Diodorus *Bibliotheca* II 9, 4), but we have no contemporary evidence for this; in any case the Persian king Xerxes is said to have damaged the ziggurat seriously and Alexander the Great is said to have repaired it (Arrian *Anabasis* 7.17.1 and Strabo *Geographica* 16.1.5), so that we cannot now form any accurate idea of its height in the last centuries BC. There were also ziggurrats at Borsippa and Uruk.

Professor Sachs's study of the texts showed that they fell into four main groups, named by him astronomical diaries (monthly or annual records of astronomical observations), goal year texts (compilations made from the diaries of previous years for the purpose of predicting astronomical events in the coming year, the goal year), almanacs (the predictions for the coming year of the length of each month, the rising and setting of each planet, the zodiac positions of the planets, eclipses, equinoxes, and the movements of Sirius), and mathematical astronomical texts. This last group contains mathematical procedures for calculating the movements of the sun, moon and the five planets then known (Mercury, Venus, Mars, Jupiter, Saturn), using the Babylonian sexagesimal number system. This system, being much easier to use than contemporary Greek mathematical notation, and being capable of handling much larger numbers or smaller fractions, allowed the Babylonians to establish values for the length of the mean lunar month and for celestial movements which were taken over by the Greek astronomers and were not significantly improved upon until the time of Copernicus. Our present system of 360 degrees in a circle, 60 minutes to the hour and 60 seconds to the minute derives directly from Babylonian systems. Two smaller groups of texts contain tables of eclipses and personal horoscopes. It is in the group labelled by Sachs astronomical diaries that Babylonian observations of comets have recently been found, including observations of Halley's Comet in the years 164 and 87 BC.

Babylonian astronomical diaries

The Greek astronomer Ptolemy (Claudius Ptolemaeus) writing *c.* AD 150 records (*Almagest* 3, 7) that the earliest astronomical observations available to him date from the time of the Babylonian king Nabonassar (Nabu-naṣir, 747–734 BC). Apart from a single group of Venus observations from the reign of Ammiṣaduqa (seventeenth century BC), we remain as ignorant as Ptolemy about earlier observations. Although contemporary astronomical texts do not survive from the time of Nabu-naṣir, a later eclipse table contains one record of a lunar eclipse that may be datable to his reign, and Ptolemy uses three eclipse observations from the reign of Mardokempad (Marduk-apla-iddina II, 721–710 BC). In the early years of the next century Babylonian astronomers were sending records of their observations to the Assyrian king. The correspondence of the Assyrian court astronomers gives us a detailed picture of contemporary astronomy and astrology in action and shows how astrology vitally affected the affairs of state. Most of our knowledge of Babylonian astrology is derived from the astrological omen texts preserved in the royal libraries of the seventh century BC at Nineveh, especially the great series *Enuma Anu Enlil*; the title 'scribe of *Enuma Anu Enlil*' given to astronomers both then and four hundred years later in Seleucid times emphasises the primary astrological aims of their work.

It was presumably at about this time that official concern for astrology led to the establishment of a regular and permanent record of astronomical events for the purpose of making more accurate predictions, the record which we know as the astronomical diaries. The earliest datable diary comes from the year 652 BC; another survives from 568–7 BC, three from the fifth century, and from 385 to 47 BC we have a long but frequently broken series of diaries. It is presumably from the earlier diaries, now mostly lost, that the astronomers of the Seleucid period were able to deduce the periods and variations of planetary and lunar motions which appear in their mathematical procedures.

A few diaries give observations for only one month, or even a single day; but the majority have a half-yearly record, for the first or last six or seven months of the year. The Babylonian year begins around the time of the spring equinox, and is made up of twelve or thirteen lunar months of twenty-nine or thirty days, each month beginning with the first sighting of the crescent of the new moon. The intercalary thirteenth month which has to be added to the year occasionally to keep the calendar in line with the natural solar year was from about 400 BC added in a regular pattern, with seven such months added in each period of nineteen years, producing a calendar which over long periods of time is only slightly less accurate than our current Gregorian calendar. The year 164–3 BC had such a thirteenth month, and the tablets translated here have the observations for months VIII to XII_2.

Before 305 BC the diaries are dated by the regnal year of the king who was on the throne at the beginning of the year. From 305 BC onwards they are dated by the Seleucid Era. Seleucus I Nicator, one of Alexander the Great's generals, was recognised as sole king in Babylon in 305 BC, but the Era named after him was post-dated by the Babylonians; so, SE 1 = 311–310 BC. In 171 BC the Parthians occupied Babylonia, and from then on texts may also be dated by the Arsacid Era (named after Arsaces I the founder of the Parthian dynasty), which had begun in 247 BC. For astronomers 1 BC is year 0, 2 BC is −1, and 100 BC is −99.

Each monthly record normally begins with a statement of whether the new moon crescent was sighted on the thirtieth or thirty-first evening of the previous month. The times in degrees (corresponding to four minutes) are given for the intervals between the rising and setting of the sun and moon at the beginning and end of the month and around full moon in the middle of the month. These time intervals help in predicting the date on which the next month will begin. The daily position of the moon when first seen each evening (in the first half of the month) or when last visible each morning (in the second half of the month) is given in relation to one of a series of thirty-one reference stars (today by convention known as Normal Stars) close to the ecliptic. For some reason there are gaps in the series around the constellations Sagittarius and Aquarius. The process by which this series of stars was selected is still quite unknown. A few other stars close to the ecliptic are sometimes used to fix the position of lunar eclipses or stationary points of the planets. Occasionally close approaches of the moon to planets or occultations are recorded.

The various time intervals around the rising and setting of sun and moon are recorded in time-degrees (UŠ) which were presumably measured by the water-clock; longer time units were the double-hour (*bēru*) and the three watches each into which both day and night were divided. The timing of lunar eclipses is often related to the passage of one of another group of stars, the Ziqpu-stars, across the zenith (*ziqpu*). The distances between the moon and the Normal Stars are given in cubits (KÙŠ) and fingers (SI) above (north) or below (south), in front (west) or behind (east). Only in the mathematical lunar ephemerides is there a concept of absolute ecliptic latitude. The cubit seems to have corresponded to 2° or 2.5°, and the finger to 5′ or 6′; but how these distances were actually measured or whether they were sometimes just estimated we do not know. Curiously enough the magnitude of lunar eclipses is also measured in fingers (totality = 12 fingers), but since the moon's apparent diameter is *c.* 30′ in this case a finger appears to be only 2.5′.

For the planets the diaries give dates of first and last visibility, stationary position and retrograde and forward motion, and conjunctions with Normal Stars; but it is clear that in many cases we are dealing not with observations but with predictions taken from almanacs and tables. It would for instance have been difficult to establish by eye the precise moment when a planet reached its stationary point. In many cases we are given both an observation and a theoretical calculation; a planet may have been quite high in the sky when first seen, so the astronomer would estimate that it should actually have been visible a day or two before; in the case of last visibility the record of the actual last sighting seems to be complemented by a calculation of first invisibility. The dates of the equinoxes and solstices and the heliacal rising and setting of Sirius also derive from theoretical schemes, not observations; the dates of eclipse possibilities are taken from a well attested eighteen-year eclipse cycle.

The monthly summaries also tell us in which zodiac signs the planets were seen. This again is not a matter of observation but of theoretical calculation. The division of the zodiac into the twelve constellations of 30° each now familiar to us seems to have taken place around 450 BC. Before that the Babylonian astronomical compendium *Mul-Apin* ('the Plough'), which dates before 1000 BC, lists eighteen constellations, 'gods standing on the path of the moon, through whose sectors the moon passes every month and whom he touches'. These constellations are all used by the Late Assyrian astronomers in the eighth century BC as reference points for planetary observations. *Mul-Apin* is our primary source for the identification of the constellations. It lists the stars in various groups or 'paths' with details of their heliacal risings and settings. Other texts give descriptions of the appearance of the constellations, and there is a very small group of drawings. At least in the zodiac belt the constellations correspond largely to the picture known from the detailed descriptions of Ptolemy, but there are some interesting differences. For instance Aries is not a ram but a 'hired worker'. The division of these constellations into twelve equal arcs of 30° was probably inspired by a theoretical division of the year into twelve months of 30 days each for which there is some evidence already in the Old Babylonian period, *c.* 1800 BC. An intermediate stage is represented by a tablet (WA 77824) of about the fifth century BC which lists the twelve months (ignoring the intercalary month) and their associated constellations, but assigns both Pleiades and Taurus to month II, both Orion and Gemini to month III and both Pegasus and Pisces to month XII. The final system of twelve constellations, of which the names are now much abbreviated, is first attested shortly before 400 BC. For some reason still not fully explained this zodiac system does not start at the First Point of Aries, 0° ecliptic longitude, but at about 355°, and this difference extends through the whole zodiac; this has to be taken into account in checking the Babylonian predictions.

The astronomical sections of the diaries also include endless details of local weather conditions, by night and day. At least one diary (WA Rm.718) refers to an earthquake, in March 251 BC, again confirming the basic accuracy of Diodorus. In addition the monthly summaries give us information on the fluctuating market prices of six commodities (barley, dates, pepper, cress, sesame, and wool), and measurements of the rise and fall of the Euphrates at Babylon, which will be useful to long term studies of early economic history and climate. The word translated by Halley as 'tempests' in his quotation from Diodorus would be better translated 'floods', and it seems that the Babylonians used this data too in the hope of predicting flood levels at Babylon. There are also occasional items of political history which are sometimes presented as rumour, 'It was heard that . . .'.

TABLE I
The Babylonian Normal Stars in 164 BC

Transliteration	Translation	Star	Longitude	Latitude
MÚL KUR šá DUR nu-nu	The bright star of the Ribbon of the Fishes	η Psc	356.77	+5.26
MÚL IGI šá SAG ḪUN	The front star of the head of the Hired man	β Ari	3.92	+8.41
MÚL ár šá SAG ḪUN	The rear star of the head of the Hired man	α Ari	7.57	+9.91
MÚL-MÚL	The Stars	η Tau	29.95	+3.83
is DA	The jaw of the Bull	α Tau	39.71	−5.62
ŠUR GIGIR šá SI	The northern . . . of the Chariot	β Tau	52.52	+5.21
ŠUR GIGIR šá ULÙ	The southern . . . of the Chariot	ζ Tau	54.73	−2.47
MÚL IGI šá še-pít MAŠ-MAŠ	The front star of the feet of the Twins	η Gem	63.43	−1.17
MÚL ár šá še-pít MAŠ-MAŠ	The rear star of the feet of the Twins	μ Gem	65.22	−1.04
MAŠ-MAŠ šá SIPA	(the star in) the Twins near the Shepherd	γ Gem	69.03	−7.00
MAŠ-MAŠ IGI	The front Twin	α Gem	80.27	+9.90
MAŠ-MAŠ ár	The rear Twin	β Gem	83.52	+6.52
MÚL IGI šá ALLA šá ULÙ	The front star of the Crab to the south	θ Cnc	95.70	−0.96
MÚL IGI šá ALLA šá SI	The front star of the Crab to the north	γ Cnc	97.53	+3.01
MÚL ár šá ALLA šá ULÙ	The rear star of the Crab to the south	δ Cnc	98.64	−0.01
SAG A	The head of the Lion	ε Leo	110.64	+9.55
LUGAL	The King	α Leo	119.92	+0.37
MÚL TUR šá 4 KÙŠ ár LUGAL	The small star 4 cubits behind the King	ρ Leo	126.34	+0.04
GIŠ-KUN A	The rump of the Lion	θ Leo	133.34	+9.65
GÌR ár šá A	The rear foot of the Lion	β Vir	146.64	+0.65
DELE šá IGI ABSIN	The lone star in front of the Barley-stalk	γ Vir	160.39	+2.97
SA_4 šá ABSIN	The bright star of the Barley-stalk	α Vir	173.82	−1.91
RÍN šá ULÙ	The southern Balance-pan	α Lib	195.05	+0.61
RÍN šá SI	The northern Balance-pan	β Lib	199.35	+8.74
MÚL MÚRUB šá SAG GÍR-TAB	The middle star of the head of the Scorpion	δ Sco	212.53	−1.72
MÚL e šá SAG GÍR-TAB	The upper star of the head of the Scorpion	β Sco	213.14	+1.28
SI_4	Lisi (a god)	α Sco	219.72	−4.29
MÚL KUR šá KIR_4 šil PA	The bright star of the tip of the arrow of Pabilsag	θ Oph	231.35	−1.55
SI MÁŠ	The horn of the Goat-fish	β Cap	273.99	+4.83
MÚL IGI šá SUḪUR-MÁŠ	The front star of the Goat-fish	γ Cap	291.63	−2.33
MÚL ár šá SUḪUR-MÁŠ	The rear star of the Goat-fish	δ Cap	393.39	−2.21

The Babylonian planets

In *Mul-Apin*	Transcription	In LBAT	Translation	Planet
MUL.SAG-ME-GAR	nēberu	múl-babbar	The ferry	Jupiter
dele-bat	delebat	dele-bat	(unknown)	Venus
UDU-IDIM-GU_4-UD	šiḫṭu	GU_4-UD	Jumping	Mercury
UDU-IDIM-SAG-UŠ	kayamānu	GENNA	Constant	Saturn
ṣal-bat-a-nu	ṣalbatānu	AN	(unknown)	Mars

TABLE I – *continued*
The Babylonian zodiacal constellations

In *Mul-Apin*	In WA 77824	In LBAT	Translation	Equivalent
MUL.LÚ.ḪUN-GÁ	MÚL.LÚ-ḪUN-GÁ	ḪUN	The hired man	Aries
MUL-MUL	MÚL-MÚL	MÚL-MÚL	The stars	Pleiades
MUL.GU$_4$-AN-NA	MÚL.GU$_4$-AN-NA	(MÚL.GU$_4$-AN-NA)	The bull of heaven	Taurus
MUL.SIPA-ZI-AN-NA	MÚL.SIPA-ZI-AN-NA	(SIPA)	The true shepherd of Anu	Orion
MUL.ŠU-GI			The old man	Perseus
MUL.ZUBI			The hooked staff	Auriga
MUL.MAŠ-TAB-BA-GAL-GAL	MÚL.MAŠ-TAB-BA-GAL-GAL	MAŠ-MAŠ	The (great) twins	Gemini
MUL.AL-LUL	MÚL.AL-LUL	ALLA	The crab	Cancer
MUL.UR-GU-LA	MÚL.UR-GU-LA	A	The lion	Leo
MUL.AB-SÍN	MÚL.AB-SÍN	ABSIN	The barley-stalk	Virgo
MUL.zi-ba-ni-tum	MÚL.zi-ba-nit	RÍN	The balance	Libra
MUL.GÍR-TAB	MÚL.GÍR-TAB	GÍR-TAB	The scorpion	Scorpius
MUL.PA-BIL-SAG	MÚL.PA-BIL-SAG	PA	Pabilsag (a god)	Sagittarius
MUL.SUḪUR-MÁŠ.KU$_6$	MÚL.SUḪUR-MÁŠ	(SUḪUR)-MÁŠ	The goat-fish	Capricornus
MUL.GU-LA	MÚL.GU-LA	GU	The giant	Aquarius
	MÚL.AŠ-GÁN		The field	Pegasus
MUL.KUN.MEŠ	MÚL.KUN.ME	zib.ME	The tails	Pisces
MUL.SIM-MAḪ			The swallow	SW Pisces
MUL.a-nu-ni-tum			Anunitum (a goddess)	NE Pisces

The Babylonian view of comets

The earliest mention of a comet in the diaries is datable to February 234 BC (LBAT 283) and may be an observation of a comet also recorded in Chinese sources. Two tablets, published as LBAT 378 and 380, contain the record of Halley's Comet in 164 BC, and one small fragment, WA 41018 (not previously published), has part of the record for 87 BC. The Babylonian word for comet, *ṣallummû*, has previously been translated 'meteor', but descriptions of the movement of a *ṣallummû* over a long period of time make it clear that at least in the diaries 'comet' is the correct interpretation. Other Babylonian words have occasionally been translated 'comet', but lack of astronomical information made the translation highly uncertain.

Although 234 BC is the earliest historical reference to a comet in the Babylonian texts, a possible earlier description of a comet survives on a tablet from the Assyrian libraries at Nineveh (WA K.250), an astrological compendium perhaps to be dated to the seventh or eighth century BC. This quotes part of an astronomical omen, 'If a star which has a coma (*ṣipru*) in front, a tail (*zibbatu*) behind, is seen, like a *ṣallummû*, like a *mišḫu* of the stars.' The word *mišḫu* seems to be used to describe a variety of luminous phenomena, including a comet's tail (as in the translation of WA 41018 below), and may in some astrological texts be an alternative word for 'comet'.

We have no precise indications of the Babylonians view of the significance of comets. All astronomical phenomena were seen as having relevance primarily to the king and the state. It is noteworthy that the only historical references included in the tables of lunar and solar eclipses were to the death and accession of kings; thus a well-known but still unpublished eclipse text (WA 32234) contains a reference to the murder of Xerxes by his son in 465 BC. Our diary for 87 BC contains an apparent reference to the accession of Gotarzes as sole king after the death of Mithradates II. This occurred in month IV, and the text also indicates that the comet was visible, perhaps for the first time, in month IV. A Seleucid king list has been interpreted as recording that the death of Antiochus IV Epiphanes was reported in month IX, 164–3 BC. His death would have been preceded by the appearance of the comet which is recorded in the historical summary for month VIII. The historical summary for month IX is missing, but the broken account of month X in WA 41670 remarkably contains a mention of the dead king's corpse; his son and successor Antiochus V is mentioned in the next line. No other reference to the corpse of a Babylonian king is known from the historical texts. Thus in both cases the appearance of Halley's Comet is recorded on tablets which also refer to the death and accession of kings. If Diodorus Siculus is correct in reporting that the Babylonians attempted to predict the occurrence of comets, then it is quite likely that they would have compiled tables similar to those which survive for eclipses, and that they would have included similar brief references to the death and accession of kings. Like the eclipse tables these would have been available to Diodorus, Ptolemy, and other historians and astrologers of the Hellenistic and Roman eras. Thus the Babylonian preoccupation with royal astrology may have played its part in the growth of the idea of the evil influences of comets.

The Babylonian diaries for 164–3 BC

The Babylonian observations of Halley's Comet are recorded in the diaries for the second half of the year SE 148 (October 164 to April 163 BC) and the first half of the year SE 225 (April to October 87 BC). Although the date would have been recorded on the tablets, all the relevant tablets are broken and the dates have been lost. The tablets therefore have to be dated by astronomical calculation using the lunar and planetary observations recorded on them. The script and the abbreviations used for the names of stars indicate a date after 400 BC. No diaries are known after 40 BC. Within these limits the astronomical data is more than adequate to establish a unique date for each fragment separately.

For 164–3 BC there are two tablets (WA 41628 = LBAT 378 and WA 41462 = LBAT 380) which each have part of the astronomical record of months VIII, IX, XII, and XII_2. A further tablet (WA 41670 = LBAT 379) fills part of the gap with the record for months X and XI, but contains no observation of the comet. It is included here both to give a more complete picture and because of its interesting references to Antiochus IV and V. The record for months I to VII would have been inscribed on a separate group of tablets which have probably not survived. It is generally assumed that the half-yearly diaries were written up shortly after the end of the final month from daily or monthly notes, of which several have survived. It is worth noting that WA 41670 has at one point in the text a note *ḫe-pí* 'broken', indicating that the text from which the scribe was copying was already damaged. Whether this means that the monthly note from which the scribe was compiling his half-yearly diary was already broken, or whether a scribe some months or years later was recopying the complete diary we cannot say.

As a further check on the date of these tablets we have three later goal year texts which excerpt information from these diaries for the purpose of predicting observations of the moon and planets at the appropriate later interval. The movements of Mercury repeat at intervals of forty-six years. WA 34034 (LBAT 1285) is a goal year text for SE 194 including observations of Mercury for SE 148. The movements of Saturn repeat at intervals of fifty-nine years. WA 35420 (LBAT 1291) is a goal year text for SE 207 including observations of Saturn for SE 148. Both of these texts partially duplicate the diaries for SE 148. The movements of the moon repeat at intervals of eighteen years. WA 34037 (LBAT 1264), the goal year text for SE 167, includes a copy of the lunar eclipse report from SE 148 month XII_2 (in case that intercalary month should parallel the first month of SE 167). It allows us to complete the fragmentary account of the eclipse preserved in the diaries. All three of these texts give additional confirmation of our dating of the diaries for 164–3 BC.

The manner in which the Babylonian observations in 164 and 87 BC relate to the orbit of the comet as calculated by Yeomans and Kiang has already been discussed in depth by Stephenson, Yau and Hunger, and the technical details and diagrams are not repeated here. In what follows we take the identification of the comets seen by the Babylonians with Halley's Comet as established. In 164 BC the comet made a close approach to the Earth and should have been particularly well placed for observation from September to November.

The observations of the comet are recorded in the monthly summaries for month VIII (October 20 to November 18) in WA 41462 and WA 41628 with

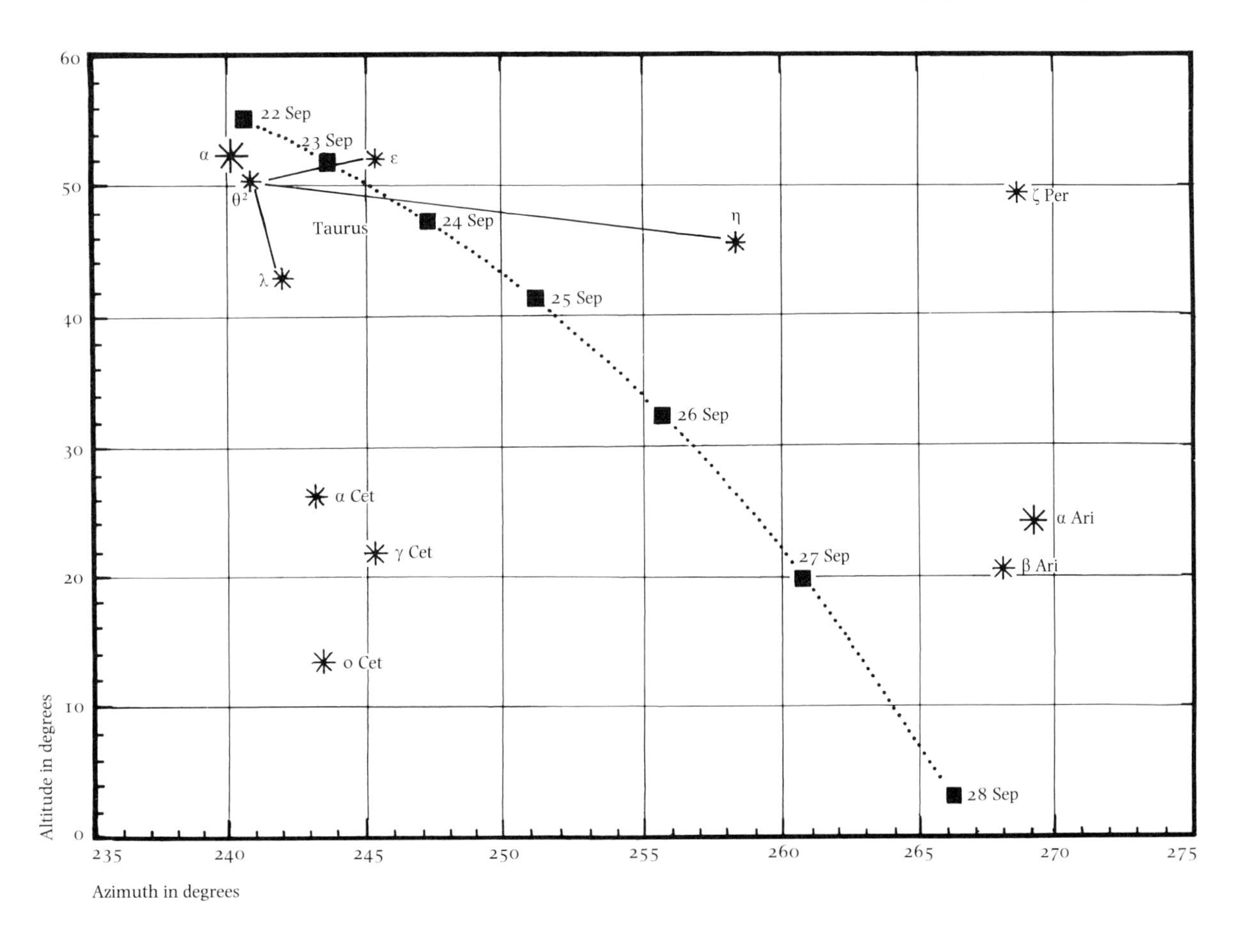

The position of Halley's Comet as seen looking west from Babylon on the mornings of 22–28 September 164 BC

slight differences. The differences suggest that we either have the comments of two different astronomers or two extracts from a longer and more detailed record. They read, 'The comet which previously had been seen in the east in the path of Anu in the area of Pleiades and Taurus, to the west . . . [. . .], and passed along in the path of Ea' (WA 41462 obverse 16–17) and '[. . . in the path] of Ea in the area of Sagittarius, 1 cubit in front of Jupiter, (the comet) being 3 cubits high to the north, . . .' (WA 41628 obverse 9').

The statement that the comet had previously been seen in the east seems to be a reference to an observation in month VI or VII, when the comet would first have become visible in the area of Taurus high in the eastern sky at about midnight. The Babylonians divided the sky into three bands or 'paths' of stars roughly parallel to the equator, of Enlil extending north from about +17°, of Anu between +17° and −17°, and of Ea south from about −17°. The Pleiades and Taurus (the constellation, not the zodiac sign) are both assigned to the path of Anu in the text *Mul-Apin*. Calculations show that for any date of perihelion between 7 and 16 November 164 BC the comet's path would have passed directly between the Pleiades (η Tauri) and the Hyades (α Tauri). After first visibility the comet would have moved rapidly westwards, and this may be reflected in the incomplete comment, 'to the west [. . .]'. WA 41462 then describes the comet as passing along in the path of Ea; the term DIB-DIB-iq (Babylonian *ītettiq*) implies regular and continuous motion. WA 41628 adds a record of an individual observation, the comet being in the region of Sagittarius 1 cubit in front of Jupiter, 3 cubits high towards the north. Sagittarius is assigned to the 'path' of Ea in *Mul-Apin*. The reference to Jupiter is very valuable as evidence for fixing the

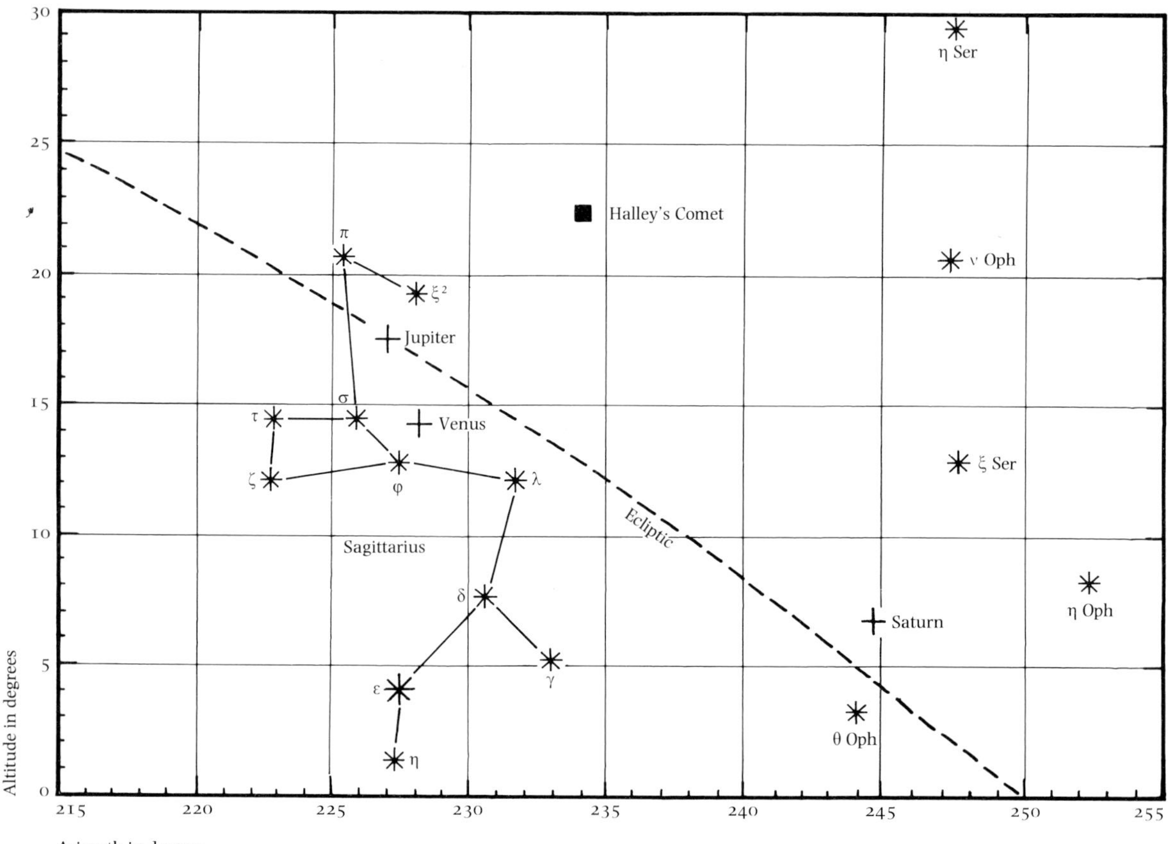

precise orbit of the comet. Comparison with other astronomical observations in the diaries suggests that the observer was recording either the closest approach of the comet to Jupiter, or perhaps more likely its position when last sighted. Jupiter was only a little east of the sun during month VIII, but the comet should

The position of Halley's Comet as seen looking west from Babylon on the evening of 26 October 164 BC

Tetradrachm of Antiochus IV Epiphanes, 175–164 BC. BMC 18

still have been visible while in the area described. However the fact that the comet was seen 1 cubit in front of Jupiter and 3 cubits to the north requires that its orbit included a date of perihelion between 9 and 26 November. Other orbits would not put the comet in this relation to Jupiter. The Babylonian observation accords with Yeomans and Kiang's estimate of November 12 as the date of perihelion. It is unfortunate that the Babylonian observation is not dated to a single day. If the comet was seen by the Babylonian astronomers after perihelion their observations would probably have been recorded in the summary for month IX, but this part of the text is still missing.

The fragmentary king-list of the Seleucid period, WA 35603, which comments on the death of each king, suggests that the death of Antiochus IV was reported at Babylon in the ninth month: [MU 1 me 48-KÁM] GAN it-te-eš-me šá ^{1}an L[UGAL . . .]. '[Year SE 148] it was heard that King An(tiochus) [died . . .].' This would have been the month after the appearance of the comet. The diary WA 41840 line 17′ contains an apparent reference to his corpse being brought back to Babylon, and in the next line his son Antiochus V is mentioned.

The Babylonian astronomical texts are written in a highly abbreviated and technical form and frequently use words taken from the earlier Sumerian language rather than their contemporary Babylonian equivalents (just as Latin is sometimes used today in legal documents). In the following transliterations Sumerian words and signs are printed in capitals. Where we have restored part of a broken text the restoration is placed between square brackets. In some cases these restorations follow the obvious sense of the text, in other cases they are based on computer calculations of the astronomical observations expected. In order to make it easier to analyse the Babylonian text we have punctuated it to match the punctuation in our translation. The original Babylonian texts do not have any form of punctuation. Month XII$_2$ is the intercalary month added after month XII; x marks an illegible sign; signs omitted by the scribe are placed within angular brackets < >; signs incorrectly added by the scribe are placed between double angular brackets ≪ ≫. The tablets are very broken and a large part of each text is missing.

Tetradrachm of Antiochus V Eupator, 164–162 BC. BMC 6

For those who wish to assess the accuracy of the Babylonian observations the table on pages 37–8 sets out side by side modern computer generated positions for the moon, stars or planets involved in each observation, and the Babylonian record of the angular distances expressed in cubits and fingers.

We calculate that in 164–3 BC the relevant Babylonian months began on the evening of the following days in the Julian calendar: Araḫsamnu (VIII) October 20, Kislīmu (IX) November 19, Ṭebētu (X) December 18, Šabāṭu (XI) January 16, Addaru (XII) February 16, Addaru atru (XII$_2$) March 17.

Babylonian tablet

WA 41462 + 41941 (LBAT 380 + 920)
SE 148, months VIII, IX, [X], [XI], XII, XII$_2$ = 164–3 BC

1 [APIN . . . DIR] ⸢AN⸣ ZA, ULÙ u KUR GIN.MEŠ. ⸢1? DIR?⸣ [. . .]
[Araḫsamnu (month VIII). . . .] cirrus [clouds], the south and east winds blew. The 1st?, clouds? [. . .]

2 [. . . GE$_6$ 3, sin e] dele-bat 2 KÙŠ, sin $\frac{1}{2}$ KÙŠ ana NIM DIB, ⸢ina IGI⸣ [múl-babbar . . .]
[. . . Night of the 3rd, the moon was above] Venus 2 cubits, the moon having passed $\frac{1}{2}$ cubit to the east, in front of [Jupiter . . .]

WA 41462, as copied by T. G. Pinches. The text describes the comet's first appearance in the area of the Pleiades and Taurus in 164 BC

3 [. . . DIR AN] ZA, ULÙ GIN. 4, kal ME DIR AN ZA, AN-B[AR$_7^{?}$. . .]
[. . .] cirrus [clouds], the south wind blew. The 4th, all day cirrus clouds, at noon$^{?}$ [. . .]

4 [. . . GE$_6$ 6, SAG GE$_6$], ⌜sin e⌝ MÚL IGI šá ⌜SUḪUR⌝ MÁŠ ⌜$2\frac{1}{2}$⌝ KÙŠ; SI GIN. 6 SI GIN. GE$_6$ 7, 7, DIR AN ZA, SI GI[N . . .]
[. . . Night of the 6th, beginning of the night], the moon was $2\frac{1}{2}$ cubits above Gamma Capricorni; the north wind blew. The 6th, the north wind blew. Night of the 7th, (and) the 7th, cirrus clouds, the north wind bl[ew . . .]

5 [. . . GE$_6$ 10 . . . , DIR A]N ZA; USAN, dele-bat SIG múl-babbar 1 KÙŠ; ina ZALÁG, ⌜GÍR Á⌝ ULÙ u KUR GÍR-GÍR. 10, ina še-rì, DIR AN Z[A$^{?}$ . . .]
[. . . Night of the 10th, . . . ,] cirrus [clouds]; first part of the night, Venus was 1 cubit below Jupiter; last part of the night, lightning flashed continuously in the south and east. The 10th, in the morning, cirrus clouds [. . .]

6 [. . . GE$_6$ 12, SA]G GE$_6$, sin ár MÚL ár šá SAG ḪUN 3 KÙŠ, sin 5 KÙŠ ana ULÙ SIG; kal GE$_6$ ŠÚ-ŠÚ; ULÙ G[IN . . .]

[. . . Night of the 12th, begin]ning of the night, the moon was 3 cubits behind Alpha Arietis, the moon being 5 cubits low to the south; all night very overcast; the south wind bl[ew . . .]

7 [. . .] SI GIN. GE_6 14, 4,30 GE_6; SI GIN; ina ZALÁG, sin ina IGI ŠUR GIGIR šá SI, 1½ KÙŠ sin 4 KÙŠ ana ULÙ [SIG . . .]
[. . .] the north wind blew. Night of the 14th, sunset to moonrise 4,30°; the north wind blew; last part of the night, the moon was 1½ cubits in front of Beta Tauri, the moon being 4 cubits [low] to the south [. . .]

8 [. . . GE_6 15, . . . GI]N?; ina ZALÁG, sin ina IGI MÚL IGI šá še-pít MAŠ-MAŠ 1 KÙŠ, sin 1½ KÙŠ ana ULÙ SIG. 15, DIR AN ZA [. . .]
[. . . Night of the 15th . . . bl]ew?; last part of the night, the moon was 1 cubit in front of Eta Geminorum, the moon being 1½ cubits low to the south. The 15th, cirrus clouds [. . .]

9 [. . . GE_6 17, ina ZALÁG sin SIG MAŠ-MAŠ ár] ⌜4?⌝½ KÙŠ, sin i-ṣa ana NIM DIB. GE_6 18, ina ZALÁG, sin ina IGI MÚL ár šá ALLA šá ULÙ 2 KÙŠ, sin 1½ K[ÙŠ ana ULÙ SIG . . .]
[. . . Night of the 17th, last part of the night, the moon was] 4?½ cubits [below Beta Geminorum], the moon having passed a little to the east. Night of the 18th, last part of the night, the moon was 2 cubits in front of Delta Cancri, the moon being 1½ cu[bits low to the south . . .]

10 [. . . GE_6 20, ina ZALÁG sin SIG LUGAL . . . KÙ]Š, sin i-ṣa ana NIM DIB. GE_6 21, ina ZALÁG, sin ár MÚL TUR šá 4 KÙŠ ár LUGAL 2½ KÙŠ, sin 1½ KÙŠ ana U[LÙ SIG . . .]
[. . . Night of the 20th, last part of the night, the moon was . . . cub]its [below Alpha Leonis], the moon having passed a little to the east. Night of the 21st, last part of the night, the moon was 2½ cubits behind Rho Leonis, the moon being 1½ cubits [low] to the so[uth . . .]

11 [. . .] x GE_6 23, SAG GE_6, DIR AN ZA; ina ZALÁG, sin ina IGI DELE šá IGI ABSIN 1 KÙŠ, sin 1½ KÙŠ ana ULÙ SIG; ina š[e-rì . . .]
[. . .] . . . Night of the 23rd, beginning of the night, cirrus clouds; last part of the night, the moon was 1 cubit in front of Gamma Virginis, the moon being 1½ cubits low to the south; in the morn[ing . . .]

12 [. . . GE_6 24, ina ZALÁG, sin IGI SA_4 šá ABSI]N 1 ⌜KÙŠ⌝, sin 8 SI ana SI NIM. 24, ina še-rì, DIR AN ZA. 25, DIR AN ZA. GE_6 26, DIR A[N . . .]
[. . . Night of the 24th, last part of the night, the moon was] 1 cubit [in front of Alpha Virgi]nis, the moon being 8 fingers high to the north. The 24th, in the morning, cirrus clouds. The 25th, cirrus clouds. Night of the 26th, clouds [. . .]

13 [. . . DIR AN] ZA. ⌜27⌝, 23,40 KUR; NU PAP. kal ME DIR AN ZA. GE_6 28, USAN, dele-bat SIG SI MÁŠ 2½ KÙŠ, x [. . .]
[. . .] cir[rus clouds]. The 27th, moonrise to sunrise 23,40°; I did not watch. All day cirrus clouds. Night of the 28th, first part of the night, Venus was 2½ cubits below Beta Capricorni, . . . [. . .]

14 [. . .] ina $MURUB_4$ ITU, 1(p) 5(b) 3 qa, ina TIL ITU, 2 PI; ZÚ-LUM, 1 GUR 3(p) ⌜4(b)⌝, ina TIL ITU, 1 GUR 3(p) 3(b); kàs [. . .]
[. . .] middle of the month, 1 *pan* 5 *sât* 3 *qa*, end of the month, 2 *pan*; dates, 1 *kur* 3 *pan* 4 *sât*, end of the month, 1 *kur* 3 *pan* 3 *sât*; *kasû*, [. . .]

15 [. . . ina] KUR SUM-in. i-nu-šú, múl-babbar ina PA, dele-bat EN $MURUB_4$ ITU ina PA ina TIL ITU ina MÁŠ, GU_4-UD ina RÍN; in 18 G[U_4-UD ina NIM ina GÍR-TAB ŠÚ . . .]
[. . .] were sold [in] the land. At that time, Jupiter was in Sagittarius, Venus was in Sagittarius until the middle of the month (and) in Capricornus at the end of the month, Mercury was in Libra; around the 18th, Mer[cury's last appearance in the east in Scorpius . . .]

16 [. . . ṣal-la]m-mu-ú šá ina IGI-ma ina ^{d}UTU-È-A ina KASKAL šu-ut ^{d}DIŠ ina KI MÚL-MÚL u MÚL.⸢GU$_4$⸣-AN-NA IGI ana ŠÚ x [. . .]
[. . . the co]met which previously had been seen in the east in the path of Anu in the area of Pleiades and Taurus, to the west . . . [. . .]

17 [. . .] x u ina KASKAL šu-ut ^{d}BE DIB-DIB-iq ITU BI U$_4$-5 6 7 SISKUR.MEŠ šá ana ^{d}EN ^{d}GAŠAN-iá u ^{d}INNIN D[Ù? . . .]
[. . .] . . . , and passed along in the path of Ea. That month, the 5th, 6th, and 7th, the sacrifices which are m[ade$^?$] to Bēl, Bēltiya and Ištar [. . .]

---

18 [GAN . . . na, mu]š; ina GUB IGI, ana šamáš NIM; sin e múl-babbar $1\frac{1}{2}$ KÙŠ, sin i-ṣa ana NIM DIB; kal GE$_6$, DIR AN [Z]A, ULÙ GIN [. . .]
[Kislīmu (month IX) . . . sunset to moonset . . . , mea]sured; (the moon) became visible while the sun stood there, it was high compared to the sun; the moon was $1\frac{1}{2}$ cubits above Jupiter, the moon having passed a little to the east; all night, cirrus clouds, the south wind blew [. . .]

19 [. . . DIR] AN ZA, ULÙ GIN. GE$_6$ 3, sin ina IGI MÚL IGI šá SUḪUR-MÁŠ 2 KÙŠ, sin $2\frac{1}{2}$ KÙŠ ana SI NIM; ⸢GÍR⸣-G[ÍR . . .]
[. . .] cirrus [clouds], the south wind blew. Night of the 3rd, the moon was 2 cubits in front of Gamma Capricorni, the moon being $2\frac{1}{2}$ cubits high to the north, lightning flas[hed . . .]

20 [. . .] ULÙ GIN. 4, DIR AN ZA, ULÙ GIN. GE$_6$ 5, SI u KUR <GIN.MEŠ>, ŠÚ$^?$-ŠÚ. 5, ŠÚ-ŠÚ; ULÙ GIN. GE$_6$ [6 . . .]
[. . .] the south wind blew. The 4th, cirrus clouds, the south wind blew. Night of the 5th, the south and east winds [blew], it was very overcast. The 5th, very overcast; the south wind blew. Night of [the 6th, . . .]

21 [. . . GE$_6$] ⸢7⸣, SAG GE$_6$, DIR AN ZA; ULÙ GIN. 7, ULÙ GIN. GE$_6$ 8, SAG GE$_6$, sin SIG MÚL KUR šá DU[R nu-nu . . .]
[. . . Night of] the 7th, beginning of the night, cirrus clouds; the south wind blew. The 7th, the south wind blew. Night of the 8th, beginning of the night, the moon was below Eta Piscium [. . .]

22 [. . . GE$_6$ 9, SAG GE$_6$, sin ár MÚL ár šá SAG ḪUN . . . K]ÙŠ, sin 5 KÙŠ ana ULÙ SIG; kal GE$_6$ DIR AN ZA. 9, DIR AN ZA; ULÙ GIN. GE$_6$ 10, S[AG GE$_6$ . . .]
[. . . Night of the 9th, beginning of the night, the moon was . . .] cubits [behind Alpha Arietis], the moon being 5 cubits low to the south; all night cirrus clouds. The 9th, cirrus clouds; the south wind blew. Night of the 10th, beg[inning of the night, . . .]

23 [. . . GE$_6$ 11, SAG GE$_6$, sin . . .] ⸢is DA⸣ 1 KÙŠ, sin 1 KÙŠ ana SI NIM; DIR AN ZA. 11, DIR AN ZA; ULÙ GIN; ina SUḪU[Š . . .]
[. . . Night of the 11th, beginning of the night, the moon was] 1 cubit [. . .] Alpha Tauri, the moon being 1 cubit high to the north; cirrus clouds. The 11th, cirrus clouds; the south wind blew; . . . [. . .]

24 [. . . GE$_6$ 12, . . .] dele-bat e MÚL IGI šá SUḪUR-MÁŠ 2 SI; ina ZALÁG, ŠÚ, ULÙ u KUR GIN.MEŠ; ina ZALÁG, AN e SA$_4$ šá [ABSIN . . .]
[. . . Night of the 12th, . . .] Venus was 2 fingers above Gamma Capricorni; last part of the night, it was overcast, the south and east winds blew; last part of the night, Mars was above Alpha Virginis [. . .]

25 [. . . GE$_6$ 13, ME . . . ,] NU ⸢PAP⸣; ULÙ u KUR GIN.MEŠ; kal GE$_6$, ŠÚ-ŠÚ; AN UTAḪ i-ṣa. 13, 2,30 na. DIR NU PAP; ŠÚ-ŠÚ; UL[Ù GIN . . .]
[. . . Night of the 13th, moonrise to sunset . . . ,] I did not watch; the south and east winds blew; all night, very overcast; a little shower. The 13th, sunrise to moonset 2,30°. (Because of) clouds, I did not watch; very overcast; the sou[th wind blew . . .]

26 [. . . UL]Ù u KUR GIN.MEŠ. EN-NUN U_4-ZAL-LA, sin TÙR NIGIN, ma-diš iq-tur_7. ina ZALÁG, sin SIG MAŠ-MAŠ IGI [. . .]
[. . . Night of the 14th, . . . the sou]th and east winds blew. During the morning watch, the moon was surrounded by a halo, it billowed very much. Last part of the night, the moon was below Alpha Geminorum [. . .]

27 [. . .] GIN. GE_6 16, DIR AN ZA; ULÙ GIN. EN-NUN U_4-ZAL-LA, ŠÚ-ŠÚ; GÍR GÍR-GÍR, GÙ U, AN [. . .]
[. . .] blew. Night of the 16th, cirrus clouds; the south wind blew. During the morning watch, very overcast; lightning flashed continuously, thunder, . . . [. . .]

28 [. . .] GE_6 17, SI GIN; ina ZALÁG, sin ina IGI LUGAL $1\frac{1}{2}$ KÙŠ, sin 2 KÙŠ ana ULÙ SIG. 17, SI GIN. x [. . .]
[. . .] Night of the 17th, the north wind blew; last part of the night, the moon was $1\frac{1}{2}$ cubits in front of Alpha Leonis, the moon being 2 cubits low to the south. The 17th, the north wind blew. [. . .]

29 [. . .] x 18, DIR AN ZA; šamáš TÙR NIGIN. ina KIN-SIG, ŠÚ. GE_6 19, SAG GE_6, DIR AN DIB x [. . .]
[. . .] . . . The 18th, cirrus clouds; the sun was surrounded by a halo. In the afternoon, overcast. Night of the 19th, beginning of the night, clouds obscured the sky . . . [. . .]

30 [. . . GE_6 2]0, ⌜kal GE_6⌝ ŠÚ-ŠÚ. ina ZALÁG, DIR AN DIB. 20, ŠÚ-⌜ŠÚ⌝ x ⌜GE_6 21⌝ x [. . .]
[. . . Night of the 2]0th, all night very overcast. Last part of the night, clouds obscured the sky. The 20th, very overcast . . . Night of the 21st . . . [. . .]

31 [. . . GE_6 26, ina ZALÁG, sin e MÚL KUR š]á KIR_4 šil PA 1 KÙŠ ⌜4? SI⌝ [. . .]
[. . . Night of the 26th, last part of the night, the moon was] 1 cubit 4 fingers [above Th]eta Ophiuchi [. . .]

Remainder of obverse lost.

Reverse (beginning lost)
(Month XII)

1′ [. . .] x [. . .]

2′ [. . .] x x ⌜GIN⌝ [DIR A]N ZA, SI GIN; AN-B[AR_7, . . .]
[. . .] . . . cirrus [clouds], the north wind blew; at noon, [. . .]

3′ [. . .] ŠÁR. ina še-rì, [. . . U]LÙ GIN; AN-BAR_7, AN ⌜dul?⌝-[. . .]
[. . . wind] gusted. In the morning, [. . .] the south wind blew; at noon, . . . [. . .]

4′ [. . .] x. GE_6 22, SI GIN. 22, dele-bat ina NIM ina zib.ME IGI; KUR NIM-a, 9 na-s[u, . . .]
[. . .] . . . Night of the 22nd, the north wind blew. The 22nd, Venus' first appearance in the east in Pisces; it was bright and high, (from) its appearance to sunrise 9°, [. . .]

5′ [. . . AN] e MÚL e šá SAG GÍR-TAB UŠ. SI GIN. ina KIN-SIG, DIR SAL AN [. . .]
[. . .] stationary point of [Mars] above Beta Scorpii. The north wind blew. In the afternoon, thin cirrus clouds [. . .]

6′ [. . .] NIM. 24 ULÙ GIN; ina KIN-SIG, DIR SAL AN ZA. GE_6 25, DIR AN ZA x [. . .]
[. . .] high?. The 24th, the south wind blew; in the afternoon, thin cirrus clouds. Night of the 25th, cirrus clouds . . . [. . .]

7′ [. . .] x ṣar-ḫu. 26, ina še-rì ŠÚ-ŠÚ, SI GIN. GE_6 27, DIR AN ZA, ULÙ G[IN . . .]

[. . .] . . . The 26th, in the morning very overcast, the north wind blew. Night of the 27th, cirrus clouds, the south wind bl[ew . . .]

8′ [. . . S]I GIN. 28, AN-BAR$_7$, DIR AN ZA; SI GIN. AN-KU$_{10}$ šamáš; ki PAP NU IGI. in [. . . GE$_6$ GIN. . . .]
[. . . the no]rth wind blew. The 28th, at noon, cirrus clouds; the north wind blew. Eclipse of the sun; I did not see it when I watched; at [. . . degrees after sunset. . . .]

9′ [. . .] qa. ZÚ-LUM, ina ⌜SAG⌝ ITU, 1 GUR 3(p) 3(b), ina MURUB$_4$ ITU, 1 GUR 3(p) 1(b), ina TIL ITU, 1 GUR 2 + [x(p) . . .]
[. . .] *qa.* Dates, at the beginning of the month, 1 *kur* 3 *pan* 3 *sât*, in the middle of the month, 1 *kur* 3 *pan* 1 *sūt*, in the end of the month, 1 *kur* 2 + [x *pan* . . .]

10′ [. . . 1]6, dele-bat ina ŠÚ ina zib.ME ŠÚ. in 19, dele-bat ina NIM ina zib.ME IGI. GU$_4$-UD ina GU. in 14, GU$_4$-U[D . . .]
[. . . The 1]6th, Venus' last appearance in the west in Pisces. Around the 19th, Venus' first appearance in the east in Pisces. Mercury was in Aquarius. Around the 14th, Mercury [. . .]

11′ [. . . ILLU . . . GI]N, PAP: 14 na; 18 19, ½ KÙŠ LAL, PAP: 17 na; TA 22 EN 26, ½ KÙŠ [. . .]
[. . . the river level ro]se [. . .], total: 14 (was the) *na* (gauge); on the 18th (and) 19th, it receded ½ cubit, total: 17 (was the) *na* (gauge); from the 22nd till the 26th, it [rose] ½ cubit [. . .]

12′ [DIR ŠE, . . . na, m]uš; DIR AN ZA; SI GIN. 1, DIR AN ZA, SI GIN. in 1, AN ana ŠÚ LAL. GE$_6$ 2, SI GIN. 2, S[I$^?$ . . .]
[Addaru atru (month XII$_2$), . . . , sunset to moonset . . . mea]sured; cirrus clouds; the north wind blew. The 1st, cirrus clouds, the north wind blew. Around the 1st, Mars moved back to the west. Night of the 2nd, the north wind blew. The 2nd, the no[rth wind . . .]

13′ [. . .] AN UTAḪ i-ṣa. 3, kal ME DIR AN DIB, AN rad PISAN MAḪ DIB, ULÙ u KUR GIN.MEŠ. G[E$_6^?$ . . .]
[. . .], a little rain shower. The 3rd, all day clouds obscured the sky, much . . . , the south and east winds blew. Ni[ght . . .]

14′ [. . . GE$_6$ 5, SAG GE$_6$, sin SIG MÚL IGI šá š]e-pít MAŠ-MAŠ 2 KÙŠ, sin ½ KÙŠ ana ŠÚ LAL; SI GIN. USAN MURUB$_4$, ŠÚ-ŠÚ. 5, DIR AN ZA; SI G[IN . . .]
[. . . Night of the 5th, beginning of the night, the moon was] 2 cubits [below Eta] Geminorum, the moon being ½ cubit back to the west; the north wind blew. First and middle part of the night, very overcast. The 5th, cirrus clouds; the north wind bl[ew . . .]

15′ [. . . TÙ]R NIGIN; SI GIN. GE$_6$ 7, SAG GE$_6$, sin ár MAŠ-MAŠ ár ⅔ KÙŠ, sin 5 KÙŠ ana ULÙ SIG; DIR AN ZA [. . .]
[. . .] was surrounded by a halo; the north wind blew. Night of the 7th, beginning of the night, the moon was ⅔ cubit behind Beta Geminorum, the moon being 5 cubits low to the south; cirrus clouds [. . .]

16′ [. . .] GIN.MEŠ. 8, ŠÚ-ŠÚ; ULÙ u KUR GIN.MEŠ; AN UTAḪ i-ṣa. GE$_6$ 9, SAG GE$_6$, sin SIG SAG A 6½ KÙŠ [. . .]
[. . . winds] blew. The 8th, very overcast; the south and east winds blew; a little rain shower. Night of the 9th, beginning of the night, the moon was 6½ cubits below Epsilon Leonis [. . .]

17′ [. . . GE$_6$ 10, SAG GE$_6$, sin ina IGI MÚL TUR šá 4 KÙŠ ár LUGAL . . . , s]in 2 KÙŠ ana ULÙ SIG. DIR AN ZA, ULÙ GIN. 10, DIR AN ZA, ULÙ GIN. GE$_6$ 11, SAG GE$_6$, si[n . . .]

[. . . Night of the 10th, beginning of the night, the moon was . . . in front of Rho Leonis], the moon being 2 cubits low to the south. Cirrus clouds, the south wind blew. The 10th, cirrus clouds, the south wind blew. Night of the 11th, beginning of the night, the mo[on . . .]

18′ [. . .] x AN dul-ḫat, IM-LÍMMU-BA GIN.MEŠ. GE_6 12, SAG GE_6, DIR AN DIB; GÍR-GÍR, GÙ U né-ḫi, AN U[TAḪ . . .]

[. . .] . . . , all four winds blew. Night of the 12th, beginning of the night, clouds obscured the sky; lightning flashed, thunder, rain sh[ower . . .]

19′ [. . .] x GIN. 13, ŠÚ-ŠÚ, AN UTAḪ i-ṣa; SI šá PA MAR šá GIN. GE_6 14, SAG GE_6, sin ina IGI S[A_4 šá ABSIN . . .]

[. . .] . . . blew. The 13th, very overcast, a little rain shower; the north wind which . . . blew. Night of the 14th, beginning of the night, the moon was in front of Alpha [Virginis . . .]

20′ [. . . ME . . . ,] DIR muš. kal GE_6 DIR AN ZA. 3 UŠ ár MÚL.na-ad-dul ár ziq-pi, d[sin AN-KU_{10} . . .]

[. . . Night of the 15th, moonrise to sunset . . . ,] measured (despite) clouds. All night cirrus clouds. (When the point) 3° behind Upsilon Bootis culminated, [lunar eclipse . . .]

21′ [. . . TÚ]G AN-e GAR-in; ina AN-KU_{10} šú{ŠÚ?}, SI šá PA MAR šá GIN .ina AN-KU_{10} šú [. . .]

[. . . the 'garment?'] of the sky' was there; in this eclipse, the north wind which . . . blew. In this eclipse [. . .]

22′ [. . .] 15, 5,30 na; DIR muš. DIR AN ZA, SI GIN. GE_6 ⌜16⌝, 8,30 GE_6 [. . .]

[. . .] The 15th, sunrise to moonset: 5,30°; measured (despite) clouds. Cirrus clouds, the north wind blew. Night of the 16th, sunset to moonrise: 8,30° [. . .]

23′ [. . . GE_6 17, ina ZALÁG, sin e MÚL e šá SA]G ⌜GÍR-TAB⌝ $\frac{1}{2}$ KÙŠ, sin i-ṣa ana NIM DIB, ár AN 1 KÙŠ 4 [SI; . . .]

[. . . Night of the 17th, last part of the night, the moon was] $\frac{1}{2}$ cubit [above Beta] Scorpii, the moon having passed a little to the east, behind Mars 1 cubit 4 [fingers; . . .]

24′ [. . . GE_6 18, ina ZALÁG, sin ár] ⌜SI_4?⌝ 2 KÙŠ, sin ⌜1⌝ KÙŠ ana SI NIM; x [. . .]

[. . . Night of the 18th, last part of the night, the moon was] 2 cubits [behind] Alpha Scorpii, the moon being 1 cubit high to the north; . . . [. . .]

25′ [. . . D]IR AN [ZA], ULÙ GIN. 20 ina še-rì SI [GIN . . .]

[. . . cirrus] clouds, the south wind blew. The 20th, in the morning, the north wind [blew . . .]

26′ [. . . T]ÙR? NIGIN, KÁ-šú ana ULÙ BE. x [. . .]

[. . . the moon] was surrounded by a halo, its door was open to the south . . . [. . .]

27′ [. . . GE_6 23, ina ZALÁG, sin ár MÚL IGI] šá SUḪUR-MÁŠ 1$\frac{1}{2}$ KÙŠ sin 2$\frac{1}{2}$ K[ÙŠ . . .]

[. . . Night of the 23rd, last part of the night, the moon was] 1$\frac{1}{2}$ cubits [behind Gamma] Capricorni, the moon being 2$\frac{1}{2}$ cu[bits high to the north. . . .]

28′ [. . . TÙR] NIGIN, KÁ-šú ana ULÙ BE. x [. . .]

[. . . the moon] was surrounded [by a halo], its door was open to the south. . . . [. . .]

29′ [. . . D]IR AN ZA, SI GIN. GE_6 20 + [. . .]

[. . .] cirrus clouds, the north wind blew. Night of the 20th? [. . .]

30′ [. . . D]IR AN ZA, kal ME ŠÚ [. . .]

[. . .] cirrus clouds, all day overcast [. . .]

Upper edge: illegible traces of five lines.

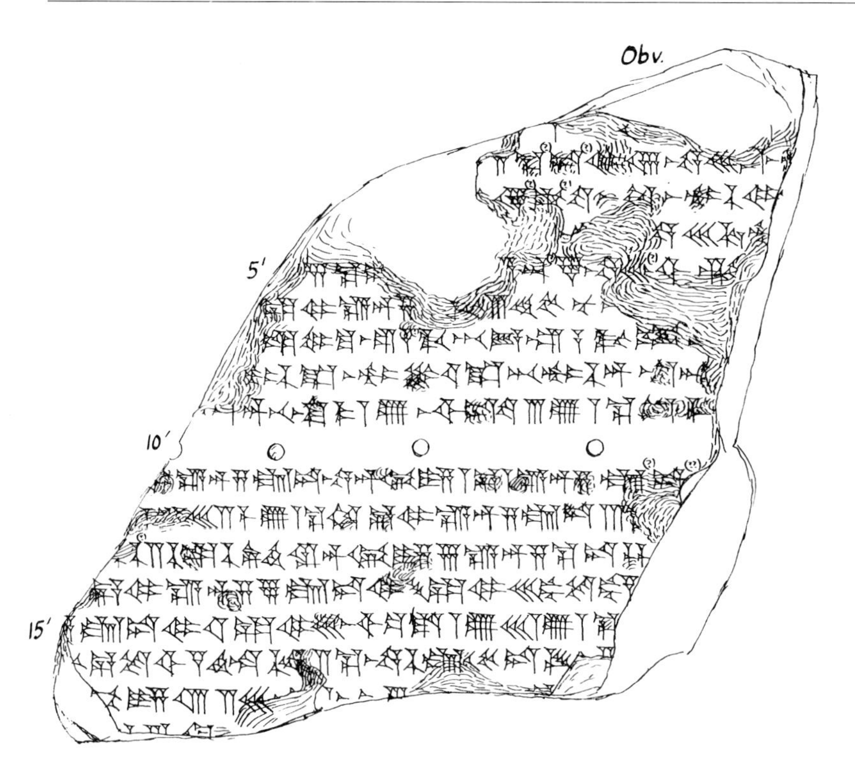

WA 41628, as copied by T. G. Pinches. The text describes the comet's position in Sagittarius close to Jupiter in 164 BC

Babylonian tablet

WA 41628 (LBAT 378)

SE 148, months VIII, IX, [X], [XI], XII, XII$_2$ = 164–3 BC

Obverse (first line lost)
(Month VIII)

1′ [. . .] x x [. . .]

2′ [. . . A]N ⸢ZA; SI GIN⸣. GE$_6$ 16, ina ZALÁG, sin ár ⸢MAŠ⸣-M[AŠ šá SIPA . . .]
[. . .] cirrus [clouds]; the north wind blew. Night of the 16th, last part of the night, the moon was behind [Gamma] Geminorum [. . .]

3′ [. . . i]n 18, GU$_4$-UD ina NIM ina GÍR-TAB ŠÚ. GE$_6$ 1[9 . . .]
[. . . Ar]ound the 18th, Mercury's last appearance in the east in Scorpius. Night of the 1[9th$^?$ . . .]

4′ [. . .] G[E$_6$ 22, ina] ZALÁG, sin SIG GÌR [ár šá A . . .]
[. . .] Night of [the 22nd], last part of the night, the moon was below Beta Vi[rginis . . .]

5′ [. . . DIR AN Z]A$^?$, SI G[IN; GE$_6$ 24, DI]R AN ZA. ina ZALÁG, sin ina IGI S[A$_4$ šá ABSIN . . .]
[. . .] cirrus [clouds], the north wind bl[ew. Night of the 24th], cirrus [cl]ouds. Last part of the night, the moon was in front of Al[pha Virginis . . .]

6′ [. . . GE$_6$ 27, S]AG GE$_6$, DIR AN ZA. ⸢27⸣, ⸢23⸣, 40 KUR; NU ⸢PAP⸣. k[al ME . . .]
[. . . Night of the 27th, be]ginning of the night, cirrus clouds. The 27th, moonrise to sunrise 23,40°; I did not watch. A[ll day . . .]

7′ [. . .] ZÚ-LUM-MA, 1 GUR 2(p) 4(b), ina TIL ITU, 1 GUR 2(p) 3(b). kàs, x [. . .]

[. . .] Dates, 1 *kur* 2 *pan* 4 *sât*, in the end of the month, 1 *kur* 2 *pan* 3 *sât*. *kasû*, [. . .]

8′ [. . . in 18 GU$_4$-UD in]a GÍR-TAB ŠÚ; GENNA ina GÍR-TAB; in 11, GENNA ina TIL GÍR-TAB ŠÚ; AN ina ABSIN; x [. . .]
[. . . around the 18th Mercury's] last appearance in Scorpius; Saturn was in Scorpius; around the 11th, Saturn's last appearance in the end of Scorpius; Mars was in Virgo; . . . [. . .]

9′ [. . . ina KASKAL š]u-⸢ut⸣ ^{d}BE ina KI PA 1 KÙŠ ina IGI múl-babbar 3 KÙŠ ana SI ⸢NIM⸣ x [. . .]
[. . . in the path] of Ea in the area of Sagittarius, 1 cubit in front of Jupiter, (the comet) being 3 cubits high to the north, . . . [. . .]

10′ [. . .] (blank) [. . .]

11′ [GAN, . . .] x; DIR AN ZA, ULÙ GIN; ina ZALÁG, AN UTAH̯ i-ṣa. 1, kal M[E D]IR AN ZA, ULÙ ⸢GIN⸣. [. . .]
[Kislīmu (month IX), . . .] . . . ; cirrus clouds, the south wind blew; last part of the night, a little rain shower. The 1st, all day cirrus clouds, the south wind blew. [. . .]

12′ [. . . GE$_6$ 2$^?$, SAG GE$_6$, sin ár dele-bat . . . ana NI]M$^?$ ⸢DU⸣, sin $2\frac{1}{2}$ KÙŠ ana SI NIM; kal GE$_6$, DIR AN ZA; ULÙ GIN. 3, D[IR . . .]
[. . . Night of the 2nd$^?$, beginning of the night, the moon] was [. . . behind Venus$^?$ to] the east, the moon being $2\frac{1}{2}$ cubits high to the north; all night, cirrus clouds; the south wind blew. The 3rd, cl[oud . . .]

13′ [. . . 1-e]n-⸢šú 2-šú⸣ GÙ-šú né-h̯i ŠUB; AN UTAH̯ i-ṣa. 6, DIR AN ZA, SI GIN, ŠÚ-ŠÚ [. . .]
[. . .] once (or) twice there was low thunder; a little rain shower. The 6th, cirrus clouds, the north wind blew, it was very overcast [. . .]

14′ [. . .] kal GE$_6$ DIR AN ZA. 8, ULÙ GIN. GE$_6$ 9, SAG GE$_6$, sin ár MÚL ár šá [SAG H̯UN . . .]
[. . .] all night cirrus clouds. The 8th, the north wind blew. Night of the 9th, beginning of the night, the moon was behind Alpha [Arietis . . .]

15′ [. . . DIR AN Z]A, ULÙ GIN. GE$_6$ 11, SAG GE$_6$, sin ina IGI is DA 1 KÙŠ, sin 1 KÙŠ ana SI [NIM . . .]
[. . .] cirrus [clouds], the south wind blew. Night of the 11th, beginning of the night, the moon was 1 cubit in front of Alpha Tauri, the moon being 1 cubit [high] to the north [. . .]

16′ [. . . GE$_6$ 12, . . . , USAN, dele-b]at e MÚL IGI šá SUH̯UR MÁŠ 2 SI. ina ZALÁG, ŠÚ; ULÙ u KUR GIN.MEŠ. ⸢ina ZALÁG⸣ [. . .]
[. . . Night of the 12th, . . . , first part of the night, Ven]us was 2 fingers above Gamma Capricorni. Last part of the night, overcast; the south and east winds blew. Last part of the night, [. . .]

17′ [. . . UT]AH̯ i-ṣa. 13, 2,30 na; ⸢DIR NU PAP⸣; ⸢ŠÚ⸣ [. . .] x [. . .]
[. . .] little [rain] shower. The 13th, sunrise to moonset, 2,30°; because of clouds I did not watch; overcast [. . .] . . . [. . .]

18′ [. . .] x x [. . .]

Remainder of obverse lost.

Reverse (beginning lost)
(Month XII)

1′ [. . .] x x [. . .] x x x [. . .]

2′ [. . . I]TU, 2(p) 1(b); ina TIL ITU, 2(p) 3(b) 3 qa; ZÚ-LUM-MA, ina SAG I[TU, . . .]
[. . . of the m]onth, 2 *pan* 1 *sūt*; at the end of the month, 2 *pan* 3 *sât* 3 *qa*; dates, at the beginning of the m[onth, . . .]

3′ [. . . dele-bat ina zib.ME] IGI; GU_4-UD ina GU; in 14, GU_4-UD ina NIM ina TIL GU ŠÚ; GENNA ina [PA. . . .]
[. . .]first appearance of [Venus in Pisces]; Mercury was in Aquarius; around the 14th, last appearance of Mercury in the east in the end of Aquarius; Saturn was in [Sagittarius. . . .]

4′ [. . .] 1? KÙŠ GIN, PAP: 20 na (blank) [. . .]
[. . . the river level] rose 1? cubit, total: the *na* (gauge) was at 20.

5′ [DIR ŠE, . . .] $AN\text{-}BAR_7$, šamáš TÙR NIGÍN, ma-diš $iq\text{-}tur_7$; ina KIN-SIG, ŠÚ. GE_6 3, kal G[E_6 . . .]
[Addaru atru (month XII_2), . . .] at noon, the sun was surrounded by a halo, it billowed very much; in the afternoon, overcast. Night of the 3rd, all night [. . .]

6′ [. . . K]IN-SIG, DIR AN ZA, SI GIN. GE_6 5, SAG GE_6, sin SIG MÚL IGI šá š[e-pít MAŠ-MAŠ . . .]
[. . .] in the afternoon, cirrus clouds, the north wind blew. Night of the 5th, beginning of the night, the moon was [. . .] below Eta Gem[inorum . . .]

7′ [. . . šam]áš TÙR NIGÍN, SI GIN. GE_6 7, SAG GE_6, sin ár MAŠ-MAŠ ár $\frac{2}{3}$ K[ÙŠ . . .]
[. . . the su]n was surrounded by a halo, the north wind blew. Night of the 7th, beginning of the night, the moon was $\frac{2}{3}$ cubit behind Beta Geminorum [. . .]

8′ [. . . GI]N.MEŠ; AN UTAḪ i. GE_6 9, SAG GE_6, sin SIG SAG A $6\frac{1}{2}$ KÙŠ [. . .]
[. . . winds bl]ew; a little rain shower. Night of the 9th, beginning of the night, the moon was $6\frac{1}{2}$ cubits below Epsilon Leonis [. . .]

9′ [. . . GI]N. GE_6 11, SAG GE_6, sin SIG GIŠ.KUN A 7 KÙŠ, sin $\frac{1}{2}$ KÙŠ ana ⌜ŠÚ?⌝ [. . .]
[. . . bl]ew. Night of the 11th, beginning of the night, the moon was 7 cubits below Theta Leonis, the moon being $\frac{1}{2}$ cubit [back] to the west [. . .]

10′ [. . .] x, DIR AN ZA, SI GIN. 12, LÁL-tim; NU PAP. DIR AN ZA, SI GIN. GE_6 [. . .]
[. . .] . . . , cirrus clouds, the north wind blew. The 12th, equinox; I did not watch. Cirrus clouds, the north wind blew. Night of the [. . .]

11′ [. . .] ⌜SI GIN?⌝. 14, 4,40 ŠÚ; DIR AN ZA, SI GIN. ina KIN-SIG, ŠÚ-ŠÚ, AN UTAḪ [. . .]
[. . .] the north wind blew. The 14th, moonset to sunrise, 4,40°; cirrus clouds, the north wind blew. In the afternoon, very overcast, rain shower [. . .]

12′ [. . . T]A ULÙ ana MAR ZALÁG-ir. 20 GAR u ZALÁG. $AN\text{-}KU_{10}$ šú, TÚG AN-e [GAR . . .]
[. . .] it cleared from south to west. 20° duration of onset and clearing. <In> this eclipse, the 'garment of the sky' [was there. . . .]

13′ [. . .] GE_6 ana ZALÁG. 15, 5,30 na; DIR muš; DIR AN ZA, S[I . . .]
[. . .] before sunrise. The 15th, sunrise to moonset 5,30°; measured (despite) clouds; cirrus clouds, the nor[th wind blew . . .]

14′ [. . . GE_6 17, SAG GE_6, sin e MÚL e šá] SAG GÍR-TAB $\frac{1}{2}$ KÙŠ, sin i-ṣa ana NIM DIB; ár AN 1 KÙ[Š 4 SI . . .]
[. . . Night of the 17th, last part of the night the moon was] $\frac{1}{2}$ cubit [above Bet]a Scorpii, the moon having passed a little to the east, behind Mars 1 cubit [4 fingers . . .]

15′ [. . .] x GE_6 19, ULÙ GIN. ina ZALÁG, sin ina IGI GEN[NA . . .]
[. . .] . . . Night of the 19th, the south wind blew. Last part of the night, the moon was in front of Sat[urn . . .]

16′ [. . .] x x [D]IR AN ZA, šamáš TÙR [NIGÍN . . .]
[. . .] . . . cirrus clouds, the sun [was surrounded] by a halo [. . .]
Illegible traces of inscription on the upper edge.

Babylonian tablet

WA 41670 + 41840 + 41915 + 42239
(LBAT 379 + 891 + 911 + 993)
SE 148, months [VIII, IX], X, XI, [XII, XII$_2$] = 164–3 BC

Obverse (beginning lost)
(Month X)

1′ [. . .] x x [. . .]

2′ [. . .] ⌜SI/ULÙ?⌝ GIN. 7 SI ⌜GIN⌝ [. . .]
[. . . the nor]th/south? wind blew. The 7th, the north wind blew [. . .]

3′ [. . . ALL]A? IGI. GE$_6$ 10, SAG GE$_6$, sin ⌜ár⌝ [. . .]
[. . . The xth, . . . (planet's)] first appearance [in Canc]er?. Night of the 10th, beginning of the night, the moon was [. . .] behind [Alpha Tauri . . .]

4′ [. . .] ⌜SI/ULÙ?⌝ GIN; ina KIN-SIG, SI GIN. GE$_6$ 12, ⌜SAG⌝ G[E$_6$, . . .]
[. . . the nor]th/south? wind blew; in the afternoon, the north wind blew. Night of the 12th, beginning of the ni[ght, . . .]

5′ [. . . GE$_6$ 13, SAG GE$_6$, sin SIG MAŠ-MAŠ ár 5] $\frac{1}{2}$ KÙŠ, sin $\frac{1}{2}$ KÙŠ ana ŠÚ LAL; kal GE$_6$, DIR AN ZA, AN [. . .]
[. . . Night of the 13th, beginning of the night, the moon was 5]$\frac{1}{2}$ cubits [below Beta Geminorum], the moon being $\frac{1}{2}$ cubit back to the west; all night, cirrus clouds, . . . [. . .]

6′ [. . .] ⌜SI/ULÙ?⌝ GIN. EN-NUN U$_4$-ZAL AN dul. 14, 4 na, DIR m[uš . . .]
[. . . the nor]th/south? wind blew. During the morning watch, . . . The 14th, sunrise to moonset 4°; mea[sured] (despite) clouds [. . .]

7′ [. . . A]N? ZA, SI GIN; ina še-rì, IM-DUGUD DUGUD, NIGÍN šamáš GIM [. . .]
[. . .] cirrus [clouds], the north wind blew; in the morning, heavy fog. The disk of the sun was like [. . .]

8′ [. . . GE$_6$ 17, ina ZALÁG, sin SIG GIŠ-KUN] ⌜A⌝ 7 KÙŠ, sin i-ṣa ana NIM DIB; sin TÙR NIGIN ma-[diš iq-tur$_7$ . . .]
[. . . Night of the 17th, last part of the night, the moon was] 7 cubits [below Theta Leonis], the moon having passed a little to the east; the moon was surrounded by a halo, [it billowed] very much [. . .]

9′ [. . . GE$_6$ 19, ina ZALÁG, sin SIG DELE šá IGI ABSIN . . . KÙ]Š, sin $\frac{1}{2}$ KÙŠ ana NIM DIB. 19, múl-babbar ina MÁŠ IGI. 1[9 . . .]
[. . . Night of the 19th, last part of the night, the moon was . . . below Gamma Virginis], the moon having passed $\frac{1}{2}$ cubit to the east. The 19th, Jupiter's first appearance in Capricorn. The 1[9th . . .]

10′ [. . .] KIN-SIG, ŠÚ-ŠÚ. GE$_6$ 22, SAG GE$_6$, ŠÚ-ŠÚ; kal GE$_6$ DIR [. . .]
[. . .] in the afternoon, very overcast. Night of the 22nd, beginning of the night, very overcast; all night clouds [. . .]

11′ [. . . GE$_6$ 23,] ina ZALÁG, sin e MÚL e šá SAG GÍR-TAB $\frac{1}{2}$ KÙŠ. 23, ŠÚ-ŠÚ; ina ⌜KIN-SIG, SI GIN⌝ [. . .]
[. . . Night of the 23rd,] last part of the night, the moon was $\frac{1}{2}$ cubit above Beta Scorpii. The 23rd, very overcast; in the afternoon, the south wind blew [. . .]

12′ [. . . GE$_6$ 25, ina ZALÁG, sin ár MÚL KUR šá KIR$_4$ šil PA] ⌜4⌝ KÙŠ, sin 1 KÙŠ ana SI NIM, e GENNA 1 KÙŠ; ina ZALÁG, AN e RÍ[N šá ULÙ . . .]
[. . . Night of the 25th, last part of the night, the moon was] 4 cubits [behind Theta Ophiuchi], the moon being 1 cubit high to the north, 1 cubit above Saturn; last part of the night, Mars was above [Alpha] Librae. [. . .]

13′ [. . . Z]A?, ŠÚ-ŠÚ. 27, 22 KUR; DIR NU PAP; ŠÚ-ŠÚ, ULÙ GIN. GE$_6$ 2⌜8⌝ [. . .]
[. . .] cirrus? [clouds], very overcast. On the 27th, moonrise to sunrise

22°; (because of) clouds I did not watch; very overcast, the south wind blew. Night of the 28th [. . .]

14′ [. . . 1] GUR 3 PI, ina MURUB$_4$ ITU 1 GUR 3(p) 1(b), ina TIL ITU 1 GUR 3(p) 2(b) [. . .]
[. . . 1] *kur* 3 *pan*, in the middle of the month 1 *kur* 3 *pan* 1 *sūt*, at the end of the month 1 *kur* 3 *pan* 2 *sât* [. . .]

15′ [. . .] ina MÁŠ IGI; in 20 GU$_4$-UD ina ŠÚ ina GU ŠÚ; GENNA ina ⸢PA⸣ [. . .]
[. . .] first appearance in Capricorn; around the 20th Mercury's last appearance in the west in Aquarius; Saturn was in Sagittarius [. . .]

16′ [. . . GI]N$^{?}$, <PAP>: 20 ⸢na⸣; 24 25 26, $\frac{2}{3}$ KÙŠ LAL, P[AP . . .]
[. . . the river level ro]se [. . .], <total>: 20 (was the) *na* (gauge); on the 24th, 25th (and) 26th, it receded $\frac{2}{3}$ cubit, to[tal: . . .]

17′ [. . . it-t]i LÚ.ÚŠ šá LUGAL DU.MEŠ-ni it-ti [. . .]
[. . . the men who] came [wit]h the king's corpse, with [. . .]

18′ [. . .] x sa ad šá ^{1}an A šá ^{1}an ḫe-pí ina D[Ù$^{?}$ . . .]
[. . .] . . . of An(tiochus V) son of An(tiochus IV) ≪broken≫ in . . . [. . .]

19′ [. . .] (blank) [. . .]

Reverse

20′ [ZÍZ, . . . D]IR AN ZA, SI GIN [x x x x G]IN. GE$_6$ [. . .]
[Šabāṭu (month XI), . . .] cirrus [cl]ouds, the north wind blew [. . . bl]ew. Night of [. . .]

21′ [. . . GE$_6$ 5, SAG GE$_6$, sin SIG MÚL IGI šá SAG ḪUN . . . , sin . . .] ana ŠÚ LAL. 5, DIR AN ZA [x x G]E$_6$ 6, DIR AN ZA; x [. . .]
[. . . Night of the 5th, beginning of the night, the moon was . . . below Beta Arietis, the moon being . . .] back to the west. The 5th, cirrus clouds [. . .] Night of the 6th, cirrus clouds, . . . [. . .]

22′ [. . . GE$_6$] ⸢8⸣, SAG GE$_6$, ŠÚ-ŠÚ; ina ZALÁG, AN SIG R[ÍN š]á SI 3 KÙŠ. 8, DIR AN ⸢ZA⸣ [. . .]
[. . . Night of] the 8th, beginning of the night, very overcast; last part of the night, Mars was 3 cubits below Beta Librae. The 8th, cirrus clouds [. . .]

23′ [. . .] x, SI$^{?}$ GIN; ina ZALÁG, DIR AN ZA. 10, GU$_4$-UD ina NIM ina MÁŠ 3 KÙŠ á[r . . .]
[. . .] . . . , the north$^{?}$ wind blew; last part of the night, cirrus clouds. The 10th, Mercury in the east in Capricorn 3 cubits beh[ind . . .]

24′ [. . . A]N ZA, SI GIN. GE$_6$ 12, SAG ⸢GE$_6$⸣, sin ina IGI MÚL ár šá ALLA šá S[I . . .]
[. . .] cirrus [clouds], the north wind blew. Night of the 12th, beginning of the night, the moon was [. . .] in front of Delta Cancri [. . .]

25′ [. . . G]E$_6$ 13, DIR AN ZA, SI šá PA MAR šá GIN; ⸢ŠED$_7$⸣. GE$_6$ 14, 11 ME, DIR muš; DIR AN ZA, x [. . .]
[. . . Ni]ght of the 13th, cirrus clouds, the north wind which . . . blew; cold. Night of the 14th, moonrise to sunset, 11°, measured (despite) clouds; cirrus clouds, . . . [. . .]

26′ [. . . GE$_6$ 15, SAG GE$_6$, sin] SIG MÚL TUR šá 4 KÙŠ ár LUGAL 2 KÙŠ, sin i-ṣa ana NIM DIB; SI šá PA [. . .]
[. . . Night of the 15th, beginning of the night, the moon was] 2 cubits below Rho Leonis, the moon having passed a little to the east; the north wind which [. . .]

27′ [. . .] 16, DIR AN ZA, SI šá PA MAR šá GIN, ŠED$_7$. GE$_6$ 17, SI [. . .]
[. . .] The 16th, cirrus clouds, the north wind which . . . blew, cold. Night of the 17th, the north wind [. . .]

28′ [. . .] x x [. . . 1]9, DIR AN ZA, ⸢AN⸣ x x. 19, DIR AN ZA, SI GIN. GE$_6$ 20, DIR A[N . . .]

[. . .] . . . [. . . Night of the 1]9th, cirrus clouds, . . . The 19th, cirrus clouds, the north wind blew. Night of the 20th, clouds [. . .]

29′ [. . . GE_6 21, ina ZALÁG . . . , sin . . . KÙ]Š ana ŠÚ DU, ár AN $\frac{2}{3}$ KÙŠ ana ⌜ŠÚ⌝ D[U, x] + 4 SI ana ULÙ SIG. 21 DIR AN ZA [. . .]
[. . . Night of the 21st, last part of the night, . . . , the moon being . . . cubi]ts to the west, being $\frac{2}{3}$ cubit behind Mars to the west!, [. . .] + 4 fingers low to the south. The 21st, cirrus clouds [. . .]

30′ [. . .] x; ⌜MAR⌝ GIN; $ŠED_7$-šir; ina ZALÁG, ŠÚ. 2[2?, DIR A]N ZA; ina še-rì NA_4 TUR ŠUR, kal [. . .]
[. . .] . . . ; the west wind blew; cold; last part of the night, overcast. The 2[2nd?], cirrus [clouds]; in the morning, a little hail fell, all [day . . .]

31′ [. . . ŠE]D_7-šir. GE_6 25, DIR AN DIB, SI [šá PA MAR šá] ⌜GIN; ma⌝-diš $ŠED_7$-šir [. . .]
[. . . co]ld. Night of the 25th, clouds obscured the sky, the north wind which . . . blew; very cold [. . .]

32′ [. . .] DIR AN ZA, IM-DUGUD i-ṣa, SI šá P[A MA]R šá GIN, $ŠED_7$. GE_6 2⌜7?⌝ x [. . .]
[. . .] cirrus clouds, a little fog, the north wind which . . . blew, cold. Night of the 27th?, . . . [. . .]

33′ [. . . GE_6] 28, SI šá PA MAR šá GIN; $ŠED_7$; ina ZALÁG, [DI]R AN ZA. 28, 13, 10 KUR; D[IR . . .]
[. . . Night of] the 28th, the north wind which . . . blew; cold; last part of the night, cirrus clouds. The 28th, moonrise to sunrise 13, 10°, cl[oud . . .]

34′ [. . .] 3(b); kàs 3 GUR 1(p) 2(b); saḫ-le_{10} 3(b) 4$\frac{1}{2}$ qa; [ŠE.GI]Š.Ì EN $MURUB_4$ ITU 1(p) 1(b) [. . .]
[. . . ; dates, . . .] 3 *sât*; *kasû*, 3 *kur* 1 *pan* 2 *sât*; cress?, 3 *sât* 4$\frac{1}{2}$ *qa*; sesame, until the middle of the month 1 *pan* 1 *sūt*, [. . .]

35′ [. . . ; AN . . .] EN TIL ITU ina GÍR-TAB. ITU BI, ILLU TA 1 EN [x x] $\frac{1}{2}$ KÙŠ GIN, 17 N[A . . .]
[. . . ; Mars . . .] until the end of the month was in Scorpius. That month, the river-level from the 1st to the [. . .] rose [. . .]$\frac{1}{2}$ cubits, the *na* (gauge) was at 17. [. . .]

36′ [. . .] x LÚ za-zak-ku šá ku-um LÚ šà-tam é-sag-[íl . . .]
[. . .] . . . the survey officer who instead of the administrator of the temple Esagil [. . .]

37′ [. . . L]Ú KI-MIN-MEŠ ana LÚ KI-MIN-MEŠ SUM-'a. ITU [BI . . .]
[. . .] the aforesaid gave to the aforesaid. [That] month [. . .]

38′ [. . .] x MEŠ šá ina LÚ mu-du LUGAL ak-ka-du-ú šá x [. . .]
[. . .] the . . . among the counsellors of the King of Akkad who . . . [. . .]

39′ [ŠE, . . . GE_6 1, SAG GE_6, sin SIG MÚL KUR šá DUR nu-nu . . . KÙ]Š, sin $\frac{1}{2}$ KÙŠ ana NIM DIB, ár dele-bat 3 KÙŠ ana NIM DU; [. . .]
[Addaru (month XII) . . . Night of the first, beginning of the night, the moon was below Eta Piscium . . . cubi]ts, the moon having passed $\frac{1}{2}$ cubit to the east, being 3 cubits behind Venus to the east; [. . .]

40′ [. . .] x x x; ⌜kal⌝ ME, ŠÚ-ŠÚ, ULÙ GIN [. . .]
[. . .] . . . ; all day, very overcast, south wind blew [. . .]

41′ [. . .] x x [. . .]

Remainder of reverse missing.

Supplementary texts for 164–3 BC
Babylonian tablet

WA 34034 (LBAT 1285)
Goal year text for SE 194 = 118 BC

Obverse 20–33, Mercury observations for SE 148 = 164 BC (46 years before)

20 ⸢MU⸣ 1 me 48-kam, [1]an LUGAL; BAR 26 GU$_4$-UD ina [ŠÚ ina MÚL]-MÚL IGI, KUR, NIM-A, 16 na-su, in [x IGI];
148th year (Seleucid Era), An(tiochus IV) being king, Nisannu (month I) 26th, first visibility of Mercury in [the west in Tau]rus, bright, high, from sunset to its setting 16°, (hence) about [the . . . (theoretical) first visibility];

21 ⸢GU$_4$⸣ GE$_6$ 2 USAN GU$_4$-UD SIG ŠUR ⸢GIGIR⸣ šá SI 1$\frac{1}{2}$ ⸢KÙŠ⸣; [GE$_6$ 3 USA]N GU$_4$-UD ⸢e⸣ ŠUR GIGIR šá ⸢ULÙ⸣ [x KÙŠ]
Ayyaru (month II) night of the 2nd first part of the night Mercury below Beta Tauri 1$\frac{1}{2}$ cubits; [night of the 3rd] first part of the night Mercury above Zeta Tauri [. . . cubits];

22 ⸢GE$_6$⸣ 9 USAN GU$_4$-UD e MÚL IGI šá ⸢še⸣-pít MAŠ-MAŠ 1 KUŠ 4 ⸢SI⸣, [GU$_4$-U]D 4? U ana ŠÚ LAL; GE$_6$ 11 USAN [GU$_4$-UD]
night of the 9th first part of the night Mercury above Eta Geminorum 1 cubit 4 fingers, [Mercu]ry (being) 4? fingers? back to the west; night of the 11th first part of the night [Mercury]

23 e MÚL ár šá še-pit MAŠ-MAŠ 1 KÙŠ 4 SI; GE$_6$ 15 USAN GU$_4$-UD [e MAŠ]-⸢MAŠ⸣ šá SIPA 4$\frac{1}{2}$ KÙŠ; SI[G 3 ŠÚ šá GU$_4$-UD]
above Mu Geminorum 1 cubit 4 fingers; night of the 15th first part of the night Mercury [above Ga]mma Geminorum 4$\frac{1}{2}$ cubits; Simānu (month III), 3rd, last (theoretical) visibility of Me[rcury]

24 ina ŠÚ ina MAŠ-MAŠ, TA 1 ki PAP NU IGI; ŠU, 7 GU$_4$-UD ina NIM ina MAŠ-MAŠ IGI, ⸢KUR⸣, [x] ⸢na-su⸣, in 4 [IGI];
in the west in Gemini, from the 1st when I looked for it I did not see it; Du'ūzu (month IV), the 7th first visibility of Mercury in the east in Gemini, bright, from its rising to sunrise [x] degrees, (hence) about the 4th (theoretical) first visibility;

25 23 GU$_4$-UD ina NIM ina ALLA ŠÚ, NU PAP; KIN 4 GU$_4$-UD ina ŠÚ ina ABSIN IGI; in 11 + [x ŠÚ šá G]U$_4$-[UD]
the 23rd last visibility of Mercury in the east in Cancer, not observed; Ulūlu (month VI) the 4th first visibility of Mercury in the west in Virgo; about the 11 + [?th last visibility of Mercury]

26 ina ŠÚ ina SAG RÍN, TA 16 ki PAP NU IGI; DU$_6$ 18 GU$_4$-UD ina NIM ina SAG RÍN 1$\frac{1}{2}$ [KÙŠ ár SA$_4$ šá ABSIN]
in the west in the beginning of Libra, from the 16th when I looked for it I did not see it; Tašrītu (month VII) 18th Mercury in the east in the beginning of Libra 1$\frac{1}{2}$ cubits [behind Alpha Virginis]

27 1 KÙŠ ana SI NIM IGI, KUR, NIM-A, 17 na-su, in 14 IGI; APIN 16 12 na-[su in 19 GU$_4$-UD]
1 cubit high to the north, first visibility, bright, high, from its rising to sunrise 17°, (hence) about the 14th (theoretical) first visibility; Arahsamnu (month VIII) 16th from its rising to sunrise 12°, [(hence) about the 19th Mercury's]

28 ina NIM ina GÍR-TAB ŠÚ; AB 5 GU$_4$-UD ina ŠÚ ina MÁŠ IGI, KUR, 15,30 na-su, in 3? IGI; 2[1 ŠÚ šá GU$_4$-UD ina ŠÚ]
last visibility in the east in Scorpius; Tebētu (month X) the 5th Mercury in the west in Capricorn, first visibility, bright, from sunset to its setting

15,30°, (hence) about the 3rd (theoretical) first visibility; 2[1st, last visibility of Mercury in the west]

29 ina GU, TA 20 ki PAP NU IGI; ZÍZ 10 GU$_4$-UD ina NIM ina MÁŠ 3 KÙŠ ár SI MÁŠ? [x +]1$\frac{1}{2}$ KÙŠ [x]
in Aquarius, from the 20th when I watched did not see it; Šabāṭu (month XI) the 10th, Mercury in the east in Capricorn 3 cubits behind Beta Capricorni [. . . +]1$\frac{1}{2}$ cubits [. . .]

30 ana ⌜SI⌝ IGI, KUR, 17 na-su, in 8 IGI; GE$_6$ 25 ina ZALÁG GU$_4$-UD e⌝ MÚL IGI [šá] SUHUR-MÁŠ $\frac{1}{2}$ K[ÙŠ];
to the north, first visibility, bright, from its rising to sunrise 17°, (hence) about the 8th (theoretical) first visibility; night of the 25th last part of the night Mercury above Gamma Capricorni $\frac{1}{2}$ cubit;

31 ⌜GE$_6$ 27⌝ ina ZALÁG GU$_4$-UD e MÚL ár šá SUḪUR-MÁŠ $\frac{1}{2}$ KÙŠ GU$_4$-UD $\frac{2}{3}$ KÙŠ ana x [x] x. ŠE 17 ŠÚ šá GU$_4$-[UD ina NI]M
night of the 27th last part of the night Mercury above Delta Capricorni $\frac{1}{2}$ cubit Mercury being $\frac{2}{3}$ cubit to the . . . Addaru (month XII), 17th, last visibility of Mercury in the east

32 [ina] zib.ME, TA 14 ina TIL GU ki PAP NU IGI; DIR ŠE 23 GU$_4$-UD [ina ŠÚ] ina MÚL-MÚL IGI, KUR NIM-A
[in] Pisces, from the 14th in the end of Aquarius when I watched did not see it; Addaru atru (month XII$_2$) the 23rd Mercury [in the west] in Taurus first visibility, bright, high,

33 17 na-su, in 20 ina TIL ḪUN IGI; GE$_6$ 29 USAN GU$_4$-⌜UD⌝ [e] is? DA? 4 KÙŠ
from sunset to its setting 17°, (hence) about the 20th in the end of Aries? (theoretical) first visibility; night of the 29th first part of the night Mercury [above] Alpha Tauri 4 cubits.

Babylonian tablet

WA 35420 (LBAT 1291)

Goal year text for SE 207 = 105–4 BC

Obverse 21–24; Saturn observations for SE 148 = 164–3 BC (59 years before)

21 [MU 1 me 48-kam ^{1}an LUG]AL, GU$_4$, GE$_6$ 16 ina ZALÁG?, GENNA ana ŠÚ ina LAL-šú e MÚL KUR šá KIR$_4$ šil PA 1 KÙŠ; in 19 GENNA ana ME E-⌜A⌝, [EN] ⌜26?⌝
[148th year (Seleucid Era), An(tiochus IV) being ki]ng, Ayyaru (month II), night of the 16th last part of the night, Saturn in its retrograding to the west was 1 cubit above Theta Ophiuchi; about the 19th Saturn's acronychal rising/opposition; [about] the 26th

22 [GENNA ana] ŠÚ ki-i UŠ-a 2 ⌜KÙŠ 4?⌝ SI ina IGI MÚL KUR šá KIR$_4$ šil PA UŠ; DU$_6$, GE$_6$ 8 USAN, GENNA ana NIM ina LAL-šú e? <MUL KUR šá KIR$_4$> šil PA 1 KÙŠ 6 SI
when [Saturn] became stationary [to] the west it was stationary 2 cubits 4? fingers in front of Theta Ophiuchi; Tašrītu, (month VII) night of the 8th first part of the night, Saturn in its forward motion to the east was 1 cubit 6 fingers above Theta Ophiuchi?.

23 [APIN, . . . ŠÚ šá GENNA ina T]IL GÍR-TAB, TA ⌜11⌝ ki PAP NU IGI; GAN 21 GENNA ina SAG PA 2$\frac{1}{2}$ KÙŠ ár MÚL KUR šá KIR$_4$ šil PA IGI, KUR NIM-A
[Month VIII, . . . last visibility of Saturn in the e]nd of Scorpius, from the 11th(?) when I looked for it I did not see it; Kislīmu (month IX), 21st first visibility of Saturn in the head of Sagittarius 2$\frac{1}{2}$ cubits behind Theta Ophiuchi, it was bright and high

24 [x na-su, in x I]GI; DIR-ŠE, EN 17 GENNA ana NIM ki-i UŠ-ú 1 KÙŠ ár 4-àm IGI.MEŠ šá PA UŠ
[from sunset to its setting . . . degrees, (hence) about the . . . (theoretical)] first visibility; Addaru atru (month XII$_2$), about the 17th when Saturn became stationary to the east it was stationary 1 cubit behind the front group of four in Sagittarius (Mu Sagittarii).

The lunar eclipse on the 15th of month XII$_2$ (night of March 30/31, 163 BC), of which fragmentary records survive in WA 41628 r.12′–13′ and WA 41462 r.20′–21′, is also recorded in WA 34037 (LBAT 1264), the goal year text for SE 167 (= 145/4 BC), 3′–9′: MU 1 me ⸢48⸣-[kam ¹an LUGAL] DIRI ŠE ⸢GE$_6$ 15 4 ME⸣ 3 UŠ ár MÚL. ⸢na-dul-lum ár⸣ ziq-pi ᵈsin ⸢AN-KU$_{10}$⸣ id ULÙ ki-i ⸢TAB-ú⸣ ina 10 UŠ GE$_6$ 3 SI GAR-an in 1,25 GE$_6$ ana ZALÁG. The date given, 148th year (of the Seleucid era), An(tiochus IV) being king, provides additional confirmation of the date of the other texts.

The full description of the eclipse can be reconstructed from the three available sources as follows: GE$_6$ 15, 4 ME, DIR muš, kal GE$_6$ DIR AN ZA. 3 UŠ ár MÚL.na-ad-dul ár ziq-pi, ᵈsin AN-KU$_{10}$; id ULÙ ki-i TAB-ú ina 10 UŠ GE$_6$ 3 SI GAR-an; [ina 10 UŠ GE$_6$ T]A ULÙ ana MAR ZALÁG-ir; 20 GAR u ZALÁG. <ina> AN-KU$_{10}$-šú TÚG AN-e GAR-in; ina AN-KU$_{10}$-šú SI šá PA MAR šá GIN; ina AN-KU$_{10}$-šú [. . .] in 1,25 GE$_6$ ana ZALÁG. 'Night of the 15th, moonrise to sunset 4°, measured (despite) clouds, all night cirrus clouds. At 3° after Nu Bootis culminated, lunar eclipse; when it began on the south side, in 10° of the night it made 3 fingers; [in 10° of the night] it cleared up from south to west; 20° total duration. During its eclipse the 'garment of the sky' was there; during its eclipse the north wind . . . blew; during its eclipse [. . .]. At 85° before sunrise.' By modern calculations the eclipse would have begun at 23.10h LT and ended at 00.97h LT reaching magnitude 0.123. Sunrise would have been at about 05.9h LT.

The solar eclipse on the 28th of month XII (= 15 March 163 BC), which is recorded as not seen at Babylon (WA 41462 r.8′) is calculated to have begun at 18.13h LT, when the sun was already 1.9° below the horizon, and to have reached magnitude 0.54.

Tetradrachm of Mithradates II, 123–87 BC. BMC 3

The Babylonian diary for 87 BC

Only one regrettably small fragment of the diary for the first half of SE 225 (87–86 BC) survives, and there are no goal year texts to confirm its date. However its reference to the capital city Seleucia, founded in 274 BC sets a useful time limit, and the astronomical observations recorded on the tablet are by themselves sufficient to give an exact date.

The observation of the comet is dated to the 13th day (of month V), corresponding to 24 August. We only have the right side of the tablet, so probably two-thirds of each line is missing. What survives seems to be almost entirely a description of the comet as seen in the previous month (IV). It seems likely that the observation being reported on the evening of the 13th was either the reappearance of the comet after perihelion, or its last visibility. Yeomans and Kiang compute the date of perihelion of Halley's Comet in 87 BC as August 6. The calculated daily motion of the comet was between 1° and 6° for at least a month. This seems to explain the text's phrase, 'day beyond day one cubit'. When first visible the comet would have been fairly high in the eastern sky before dawn and its tail would have pointed north-west. After conjunction it would have been visible in the evening reaching maximum elongation from the sun of about 35° some fifteen days after perihelion; at this point its tail would have pointed north-east. Shortly afterwards

Tetradrachm, possibly of Gotarzes I, 91–81 BC. BMC 2 'Artabanus II'

TABLE II
The Babylonian observations compared with modern calculations

Conjunctions of the moon with Normal Stars

Year 148 SE = 164/163 BC

Month	Day	B/E	Star	Δλ	Δβ	J date	LT	λ star	β star	λ moon	β moon	Δλ	Δβ	Alt
VIII	(06)	(B)	γ Cap	—	−2½ c	Oct 25	18.3h	291.63°	−2.33°	290.70°	+4.34°	−0.93°	+ 6.67°	+39°
VIII	(12)	B	α Ari	+3 c	−5 c	Oct 31	18.2h	7.57°	+9.91°	15.43°	−1.55°	+7.86°	−11.46°	+20°
VIII	14	E	β Tau	−1½ c	−4 c	Nov 03	05.8h	52.52°	+5.21°	47.22°	−3.71°	−5.30°	− 8.92°	+18°
VIII	(15)	E	η Gem	−1 c	−1½ c	Nov 04	05.8h	63.43°	−1.17°	60.24°	−4.50°	−3.19°	− 3.33°	+30°
VIII	16	E	(γ Gem)	+?	?	Nov 05	05.8h	69.03°	−7.00°	72.96°	−5.05°	+3.93°	+ 1.95°	+41°
VIII	(17)	(E)	(β Gem)	+little	−4½ c?	Nov 06	05.8h	83.52°	+6.52°	85.55°	−5.37°	+2.03°	−11.89°	+51°
VIII	18	E	δ Cnc	−2 c	−1½ c	Nov 07	05.9h	98.64°	−0.01°	97.92°	−5.45°	−0.72°	− 5.44°	+60°
VIII	(20)	(E)	(α Leo)	+little	?	Nov 09	05.9h	119.92°	+0.37°	122.13°	−4.91°	+2.21°	− 4.54°	+72°
VIII	21	E	ρ Leo	+2½ c	−1½ c	Nov 10	05.9h	126.34°	+0.04°	134.15°	−4.33°	+7.81°	− 4.37°	+70°
VIII	22	E	β Vir	?	−?	Nov 11	05.9h	146.64°	+0.65°	146.16°	−3.57°	−0.48°	− 4.22°	+63°
VIII	23	E	γ Vir	−1 c	−1½ c	Nov 12	05.9h	160.39°	+2.97°	158.40°	−2.65°	−1.99°	− 5.62°	+54°
VIII	(24)	E	α Vir	(−)1 c	+8 f	Nov 13	05.9h	173.82°	−1.91°	170.79°	−1.61°	−3.03°	+ 0.30°	+44°
IX	03	<B>	γ Cap	−2 c	+2½	Nov 21	18.0h	291.63°	−2.33°	286.59°	+4.33°	−5.04°	+ 6.66°	+28°
IX	08	B	η Psc	?	−?	Nov 26	17.9h	356.77°	+5.26°	357.77°	−0.02°	+1.00°	− 5.28°	+47°
IX	09	B	(α Ari)	+?	−5 c	Nov 27	17.9h	7.57°	+9.91°	11.40°	−1.22°	+3.83°	−11.13°	+42°
IX	11	B	α Tau	−1 c	+1 c	Nov 29	17.9h	39.71°	−5.62°	38.00°	−3.41°	+0.29°	+ 2.21°	+27°
IX	(14)	E	α Gem	?	−?	Dec 03	06.1h	80.27°	+9.90°	81.29°	−5.30°	+1.02°	−15.20°	+20°
IX	17	E	α Leo	−1½ c	−2 c	Dec 06	06.2h	119.92°	+0.37°	118.00°	−4.98°	−1.92°	− 5.35°	+47°
IX	26	(E)	θ Oph	?	(+)1c 4f	Dec 15	06.2h	231.35°	−1.55°	231.55°	+3.30°	+0.20°	+ 4.85°	+18°
X	10	B	(α Tau)	+?	?	Dec 27	17.8h	39.71°	−5.62°	47.66°	−4.08°	+7.95°	+ 1.54°	+44°
X	(13)	(B)	(β Gem)	−½ c	−[4]½ c	Dec 30	17.8h	83.52°	+6.52°	85.35°	−5.41°	+1.83°	−11.93°	+17°
X	(17)	(E)	(θ Leo)	+little	+7 c	Jan 04	06.2h	133.34°	+9.65°	138.29°	−3.88°	+4.95°	−13.53°	+35°
X	(19)	E	(γ Vir)	+½ c	?	Jan 06	06.2h	160.39°	+2.97°	162.19°	−2.08°	+1.80°	− 5.05°	+47°
X	(23)	E	β Sco	—	+½ c	Jan 10	06.2h	213.14°	+1.28°	212.09°	+2.11°	−1.05°	+ 0.83°	+46°
X	(25)	(E)	(θ Oph)	(+)4 c	+1 c	Jan 12	06.2h	231.35°	−1.55°	239.32°	+3.74°	−7.97°	+ 5.29°	+31°
XI	(05)	(B)	(β Ari)	−(little)	?	Jan 20	17.9h	3.92°	+8.41°	3.84°	−0.82°	−0.08°	− 9.23°	+50°
XI	12	B	δ Cnc	−?	?	Jan 27	18.0h	98.64°	−0.01°	94.59°	−5.37°	−4.05°	− 5.36°	+37°
XI	(15)	(B)	ρ Leo	+little	−2 c	Jan 30	18.0h	126.34°	+0.04°	130.57°	−4.06°	−4.23°	− 4.10°	+ 6°
XI	(21)	E	(δ Sco)	−?	?	Feb 06	05.9h	212.53°	−1.72°	207.80°	+1.95°	−4.73°	+ 3.67°	+45°
XII	01	B	η Psc	+½ c	?	Feb 16	18.2h	356.77°	+5.26°	357.94°	−0.35°	+1.17°	+ 5.61°	+23°
XII/2	05	B	η Gem	−½ c	−2 c	Mar 20	18.8h	63.43°	−1.17°	61.90°	−5.03°	−1.53°	− 3.86°	+53°
XII/2	07	B	β Gem	+⅔ c	−5 c	Mar 22	18.8h	83.52°	+6.52°	87.46°	−5.52°	−3.94°	−12.04°	+72°
XII/2	09	B	ε Leo	?	−6½ c	Mar 24	18.8h	110.64°	+9.55°	112.00°	−5.02°	+1.36°	−14.57°	+71°
XII/2	(10)	(B)	(ρ Leo)	?	−2 c	Mar 25	18.8h	126.34°	+0.04°	124.07°	−4.45°	−2.27°	− 4.49°	+64°
XII/2	11	B	θ Leo	−½ c	−7 c	Mar 26	18.9h	133.34°	+9.65°	136.05°	−3.69°	+2.71°	−13.34°	+54°
XII/2	14	B	α Vir	−?	?	Mar 29	18.9h	173.82°	−1.91°	172.16°	−0.67°	−1.66°	+ 1.24°	+23°
XII/2	(17)	(E)	(β)Sco	+little	(+)½ c	Apr 02	05.0h	213.14°	+1.28°	213.41°	+2.42°	+0.27°	+ 1.14°	+25°
XII/2	(18)	(E)	α Sco	+2 c	+1 c	Apr 03	05.0h	219.72°	−4.29°	226.37°	+3.28°	+6.65°	+ 7.57°	+30°
XII/2	(23)	(E)	(γ)Cap	(+)1½ c	(+)2½ c	Apr 08	05.0h	291.63°	−2.33°	294.91°	+3.88°	+3.28°	+ 6.21°	+33°

Year 225 SE = 87/86 BC

Month	Day	B/E	Star	Δλ	Δβ	J date	LT	λ star	β star	λ moon	β moon	Δλ	Δβ	Alt
II	(25)	E	η Tau	+2 c	−4 c	Jun 09	04.1h	31.00°	+3.84°	35.50°	−5.52°	+4.50°	− 9.35°	+12°
V	04	B	β Lib	−1 c	−1½ c	Aug 14	19.5h	200.40°	+8.73°	199.17°	+3.80°	−1.23°	− 4.93°	+44°
V	19	E	α Tau	−?	?	Aug 30	04.7h	40.77°	−5.61°	35.07°	−5.41°	−5.70°	+ 0.20°	+64°

Conjunctions of the moon with planets

Year 148 SE = 164/163 BC

Month	Day	B/E	Planet	Δλ	Δβ	J date	LT	λ planet	β planet	λ moon	β moon	Δλ	Δβ	Alt
VIII	(03)	(B)	Venus	+½ c	+2 c	Oct 22	18.4h	245.55°	−1.99°	248.06°	+3.91°	+2.51°	+ 5.90°	+18°
VIII	(03)	(B)	Jupiter	+?	?	Oct 22	18.4h	252.34°	−0.03°	248.06°	+3.91°	−4.28°	+ 3.04°	+18°
IX	01	B	Jupiter	−little	+1½ c	Nov 19	18.0h	258.27°	−0.08°	257.35°	+4.20°	−0.92°	+ 4.28°	+ 9°
X	(25)	(E)	Saturn	—	+1 c	Jan 12	06.2h	241.41°	+1.47°	239.32°	+3.74°	−2.09°	+ 2.27°	+31°
XI	(21)	(E)	Mars	+⅔ c	−4 f	Feb 06	05.9h	205.16°	+2.12°	207.80°	+1.95°	+2.64°	− 0.17°	+45°
XII	(01)	(B)	Venus	+3 c	?	Feb 16	18.2h	351.51°	+7.09°	357.94°	−0.35°	+6.43°	− 7.44°	+23°
XII/2	(17)	(E)	Mars	+1 c 4 f	?	Apr 02	05.0h	210.87°	+1.47°	213.41°	+2.42°	+2.54°	+ 0.95°	+25°
XII/2	(19)	(E)	Saturn	−?	?	Apr 04	05.0h	244.32°	+1.57°	239.80°	+3.95°	−4.52°	+ 2.38°	+35°

Year 225 SE = 87/86 BC

Month	Day	B/E	Planet	Δλ	Δβ	J date	LT	λ planet	β planet	λ moon	β moon	Δλ	Δβ	Alt
V	03	B	Venus	+2 c	?	Aug 13	19.5h	180.61°	−0.66°	184.51°	+2.97°	+3.90°	+ 3.63°	+20°

TABLE II – *continued*
The Babylonian observations compared with modern calculations

Conjunctions of planets with planets and stars

Year 148 SE = 164/163 BC

Month	Day	B/E	Star	Δλ	Δβ	J date	LT	λ star	β star	λ planet	β planet	Δλ	Δβ
VIII	(10)	F	♀ / ♃	–	−1 c	Oct 29	18.5h	♃253.74°	−0.05°	♀253.94°	−2.20°	−0.20°	−2.15°
VIII	28	F	♀ / β Cap	?	−2½ c	Nov 16	18.0h	273.99°	+4.83°	275.22°	−2.41°	+1.23°	−7.24°
IX	(12)	(F)	♀ / γ Cap	–	+2 f	Nov 30	18.0h	291.63°	−2.33°	291.37°	−2.16°	−0.26°	−0.17°
IX	(12)	E	♂ / α Vir	?	+?	Dec 01	06.0h	173.82°	−1.91°	173.33°	+1.80°	−0.49°	+3.71°
X	(25)	E	♂ / [α] Lib	?	+?	Jan 12	06.0h	195.05°	+0.61°	194.92°	+2.05°	−0.13°	+1.44°
XI	08	E	♂ / β Lib	–	−3 c	Jan 23	06.0h	199.35°	+8.74°	199.79°	+2.10°	+0.44°	−6.64°
XI	10	E	☿ / β Cap	+3 c	?	Jan 26	06.0h	273.99°	+4.83°	282.54°	+2.59°	+8.55°	−2.24°
XII	(23)	–	♂ / β Sco	–	+?	Mar 10	05.5h	213.14°	+1.28°	212.35°	+1.99°	−0.79°	+0.71°

Year 225 SE = 87/86 BC

Month	Day	B/E	Star	Δλ	Δβ	J date	LT	λ star	β star	λ planet	β planet	Δλ	Δβ
II	(25)	(B)	☿ / δ Cnc	?	?	Jun 08	20.0h	99.70°	−0.01°	99.10°	+1.07°	−0.60°	+1.08°
V	08	E	♂ / δ Cnc	?	+6 f	Aug 19	05.0h	99.70°	−0.01°	99.62°	+0.91°	−0.08°	+0.92°
V	17	B	♀ / α Lib	?	−1 c	Aug 27	19.5h	196.10°	+0.60°	196.22°	−1.74°	+0.12°	−2.34°

Additional data for year 148 SE = 164/163 BC
Computed solstice 164 BC Dec 24 08h; computed equinox 163 BC Mar 24 14h
Solar eclipse of Month XII day 28 = Mar 15, 163 BC (JDN 1661961); μ = 0.540
Beginning of eclipse ET 18.70h LT 18.13h Alt − 1.90°
Maximum phase ET 19.49h LT 18.91h Alt −13.35°
Lunar eclipse of Month XII/2 day 15 = Mar 30/31, 163 BC (JDN 1661976); μ = 0.123
Beginning of eclipse ET 23.70h LT 23.10h Alt 53.10°
End of eclipse ET 01.57h LT 00.97h Alt 52.60°

B = beginning of the night, E = end of the night, F = first watch of the night. c = cubit, f = finger.
J date = date in the Julian calendar. LT = local apparent (sundial) time; ET = ephemeris time.
Alt = altitude. λ = geocentric longitude, β = geocentric latitude, Δ = difference, μ = eclipse magnitude.
☿ = Mercury, ♀ = Venus, ♂ = Mars, ♃ = Jupiter. For lunar observations at the beginning or end of the night assumed solar depression = −10°. Lunar observations are calculated to the nearest 0.1h, but planetary observations to the nearest 0.5h. The computer programmes for these calculations were devised by Dr F. R. Stephenson and run on a GEC 4080 computer.

the comet would again have been lost to view in a second conjunction with the sun. So the description of the tail as pointing north-west and having a length of 4 cubits (8°–10°) seems to be part of the observation in month IV, before perihelion. Detailed consideration of visibility conditions leads to the conclusion that the observation on the 13th of month V was in fact of the comet's last sighting, the description of which at the beginning of line 11′ of the reverse of the tablet is still missing.

The mention of someone sitting himself on his throne (reverse line 4′) presumably refers to the accession of Gotarzes as sole king after the death of Mithradates II.

The computer calculations of the lunar and planetary observations in this text are set out on pages 37–8. We calculate that in 87 BC the relevant Babylonian months began on the evening of the following days in the Julian calendar: Nisannu (I) April 15, Ayyaru (II) May 15, Simānu (III) June 13, Du'ūzu (IV) July 13, Abu (V) August 11, Ulūlu (VI) September 9.

Babylonian tablet

WA 41018. SE 225, months [I], II, [III], IV, V, [VI] = 87 BC

Obverse (beginning lost)
(Month II)

1′ [. . . M]ÚL ár šá [. . .]
[. . .] the rear star of [. . .]

2′ [. . . GE_6 25, ina ZALÁG, sin e MÚL á]r ⌜šá⌝ ALLA šá ⌜ULÙ⌝ [. . .]
[. . . Night of the 25th, last part of the night, Mercury was above] Delta Cancri [. . .]

3′ [. . . ina ZALÁG, sin] ⌜ár⌝ MÚL-MÚL 2 KÙŠ, sin ⌜4⌝ [KÙŠ ana ULÙ SIG]
[. . . the moon was] 2 cubits behind Eta Tauri, the moon being 4 [cubits low to the south.]

4′ [. . .] PI 4 qa, TA 5 EN 9 [. . .]
[. . . x] *pan*, 4 *qa*, from the 5th to the 9th [. . .]

5′ [. . . kà]s-si-i, 1 GUR; saḫ-le_{10}, 2(b) [. . .]
[. . .] *kasû*, 1 *kur*; cress, 2 *sât* [. . .]

6′ [. . . in 24 múl-babbar ina MAŠ-M]AŠ ŠÚ; in 11 dele-bat ALLA KUR; in [. . .]
[. . . Around the 24th, Jupiter's] last appearance [in Gemi]ni; around the 11th, Venus reached Cancer; around [. . .]

7′ [. . .] x ŠÚ? GIŠ? NE? LÚ.UN.MEŠ KUR.ZI? [. . .]
[. . .] . . . The people of . . . [. . .]

8′ [. . .] LÚ šá ana muḫ-ḫi DÙ-uš NÍG.[ŠID . . .]
[. . .] the chief accountant [. . .]

9′ [. . .] x x ERÍN? u LÚ.GAL UNKIN ⌜KUR⌝ U[RI? . . . U]RU se-lu-ke-'a-a
[. . .] . . . army and the satrap of Baby[lonia . . . to] Seleucia

10′ [. . .] x-ni-šú U_4-18 LÚ.GA[L . . .]
[. . .] . . . On the 18th, the chief [. . .]

11′ [. . .] x x [. . .]

WA 41018. Babylonian observations of the comet in 87 BC

Reverse (beginning lost)
(Month IV)

1′ [. . .] x [. . .]

2′ [. . .] ⌜in⌝ ⌜27 GU_4-UD⌝ A KUR in [. . .]
[. . .] Around the 27th Mercury reached Leo. Around [. . .]

3′ [. . .]-qa-an ina a-ḫu-ul-la-a
[. . .] . . . on the other bank

4′ [. . . G]IŠ.GU-ZA-šú ú-šib
[. . .] sat himself on his throne.

5′ [NE, . . .] x ZI-IR. GE_6 3, sin ár dele-bat 2 KÙŠ
[Abu (month V), . . .] . . . Night of the 3rd, the moon was 2 cubits behind Venus.

6′ [. . . GE 4, SAG GE_6, sin ina IGI RÍN šá SI] 1 KÙŠ, sin $1\frac{1}{2}$ KÙŠ ana ULÙ SIG. in 4, GU_4-UD ina NIM ina A ŠÚ
[. . . Night of the 4th, beginning of the night, the moon was] 1 cubit [in front of Beta Librae], the moon being $1\frac{1}{2}$ cubits low to the south. Around the 4th, Mercury's last appearance in the east in Leo.

7′ [. . . GE_6 8, . . .] ZI-IR; ina ZALÁG, AN e MÚL ár šá ALLA šá ULÙ 6 SI
[. . . Night of the 8th] . . . ; last part of the night, Mars was 6 fingers above Delta Cancri

8′ [. . . GE_6] ⌜13⌝, 8 ME muš. USAN, ᵈṣal-lam-mu-ú
[. . . Night of] the 13th, moonrise to sunset 8° measured. First part of the night, the comet

9′ [. . .] šá ITU.ŠU u_4-mu al-la u_4-mu 1 KÙŠ
[. . .] which (in) month IV day beyond day 1 cubit

10′ [. . .] x bi-rit SI u MAR mi-ši-iḫ-šú 4 KÙŠ
[. . .] . . . between north and west its tail 4 cubits

11′ [. . . GI]N? 13, 2,30 ŠÚ, muš. AN-KU_{10} sin šá DIB. in ⌜1 DANNA ME⌝ N[IM-A]
[. . . wind bl]ew. The 13th, moonset to sunrise 2,30°, measured. Eclipse of the moon which was omitted; at 1 *bēru* (= 30°) after sunrise.

12′ [. . . GE_6 17?, U]SAN, dele-bat SIG RÍN šá ULÙ 1 KÙŠ; ina ZALÁG, x [. . .]
[. . . Night of the 17th, fi[rst part of the night, Venus was 1 cubit below Alpha Librae; last part of the night, . . . [. . .]

13′ [. . .] x x GE_6 19, ina ZALÁG, sin ina [IGI is DA . . .]
[. . .] . . . Night of the 19th, last part of the night, the moon was in [front of Alpha Tauri . . .]

14′ [. . .] x [. . .]

CHAPTER 3

Early Chinese observations of Halley's Comet

The period in Chinese history which parallels the Babylonian observations discussed in Chapter 2 is known as the Former Han Dynasty. This first phase of the Han Dynasty, which lasted from 206 BC to AD 9, is the earliest period from which a wide range of astronomical records have survived. A particularly fine report of Halley's Comet at its return in 12 BC is found in the *Han-shu*, the official history of the period. In this chapter we shall first summarise the extant Chinese astronomical observations of various kinds – eclipses, lunar and planetary phenomena, etc., as well as comets – from earliest times down to the end of the Former Han Dynasty, making special reference to the reliability of cometary records. After this, attention will be focused on references to Halley's Comet. We shall only consider Chinese observations. Although Korea and Japan were cultural satellites of China, at this early period there is little in the way of true history from these countries.

Summary of early Chinese history

The earliest historical dynasty in China is the Shang (roughly 1700–1050 BC). During the past century numerous original records, in the form of 'Oracle Bones' have come to light. The British Library has a substantial collection of these inscribed bones, which are of oxen and other animals, as well as numerous turtle shells. Many are mere fragments, although there are some well preserved specimens. Several hundred thousand oracle bones have been found at the village of Hsiao-t'un near An-yang in Ho-nan province. All probably date from between 1200 and 1050 BC, the closing stages of the Shang Dynasty. The Shang people practised divination on a large scale, and the bones which have been excavated carry a wide variety of questions put to the ancestral spirits. Most are very basic, e.g. 'Will the weather be fine tomorrow?' By applying a hot needle to a bone or turtle shell and studying the form of the cracks the Shang diviners interpreted the reply to their queries. Because these texts deal with oracles, few astronomical observations are preserved. However there are five known records of lunar eclipses and possibly a solar eclipse and a new star. The dates of all of these events are still in doubt. Nothing which might relate to a comet has so far been identified.

The Shang Dynasty was overthrown around 1050 BC. The new dynasty – the Chou – was by far the longest in Chinese history. However by the eighth century BC it was already becoming weak, gradually fragmenting into independent states. None the less the dynasty survived in one form or another through most of the Chan-kuo or Warring States Period until 256 BC, when the last Chou king abdicated in favour of the ruler of the state of Ch'in. This state eventually unified the whole of China in 221 BC, with Ch'in Shih-huang as its first Emperor. The array of life size models of soldiers and horses found at Ch'in Shih-huang's tomb near the city of Hsi-an has attracted much attention recently.

The history of the Chou Dynasty is rather sparse, a result of the notorious 'Burning of the Books' which took place in 213 BC. This systematic destruction of historical and other documents was ordered by Ch'in Shih-huang. His motive was to destroy the power of the feudal princes by severing links with the past. To this end the purge was very successful but the harsh Ch'in regime lasted only another seven years. Sadly, when the Ch'in capital of Hsien-yang was looted by rebels in 206 BC, surviving copies of the prescribed literature perished in the flames.

Only a single detailed historical work is known to originate from the Chou Dynasty. This work, the *Ch'un-ch'iu* ('Spring and Autumn Annals') now only exists in late copies. It is a chronicle of Lu, one of the States into which China was divided at this time. It covers the period from 722 to 481 BC and there is a distinct possibility that the book was edited by Confucius. Many of the details in the *Ch'un-ch'iu* are elaborated in an early commentary, the *Tso-chu'an*, compiled during the Warring States period. The astronomical records in the *Ch'un-ch'iu* consist mainly of solar eclipse reports. However there are also a few references to comets and meteors.

Any other state chronicles apart from that of Lu are now lost. The only remaining histories of the period (apart from the *Ch'un-ch'iu*) are the *Shih-chi* ('Historical Records') and the 'Bamboo Annals'. Both of these are wide ranging but are often lacking in detail. The more reliable and informative is the *Shih-chi*, compiled by the Grand Historian of China, Szu-ma Ch'ien, between 104 and 87 BC. This is a history of China from earliest (legendary) times to the writer's own time (actually 122 BC). Although detailed in later periods, much of the earlier material in the *Shih-chi* is very brief. In particular, there are no astronomical records before the period covered by the *Ch'un-ch'iu*, and over the next 240 years there is nothing independent of this chronicle. However from 480 BC down to the beginning of the Former Han Dynasty the *Shih-chi* cites a number of observations, mainly of eclipses and comets, which are not found in any other early work. Most were apparently taken from a chronicle of the state of Ch'in which is now lost.

The *Chu-shu Chi-nien* or 'Bamboo Annals' is a brief chronicle of events from legendary times to about 300 BC. It tends to be rather sporadic, often omitting several years at a time. The annals are on record as having been discovered in AD 279 in the tomb of a ruler of Wei – one of the Warring States – who had died soon after the annals were completed. Such circumstances were formerly regarded as rather dubious and doubts were expressed regarding the reliability of the information in the chronicle. However, in view of the exciting literary and other finds made recently in tombs of comparable age such criticisms now seem less viable. Nevertheless, as a source of astronomical records the *Chu-shu Chi-nien* is disappointing. As far back as 720 BC there is virtually nothing of interest which is independent of the *Ch'un-ch'iu*. Around the middle of the tenth century BC there is an allusion to a comet, but it is difficult to assess the reliability of such an early and isolated record.

Finally in our brief survey of ancient Chinese history we come to the Former Han Dynasty. This was founded by Liu Pang, who successfully led a rebellion against the Ch'in regime. Although the last Ch'in emperor was killed in 206 BC, it was not until four years of turmoil had elapsed that Liu Pang was proclaimed emperor. The new dynasty, known as the Han, proved to be largely a time of peace – something China had not known for many centuries. In its first phase (the Former or Western Han), it lasted for more than two centuries. In AD 9 the throne was usurped by Wang Mang, but his reign proved disastrous. When he was killed by rebels in AD 23 much of his capital city – the Han metropolis of Ch'ang-an – was devastated. No doubt many valuable historical documents perished in the flames. The Han Dynasty was now re-established, but with its capital at Lo-yang, far to the east of the ruined city of Ch'ang-an. This later phase of the dynasty, which lasted until AD 220, is thus often known as the Eastern Han as well as the Later Han. The official history of the Former Han Dynasty, the *Han-shu*, was compiled by Pan Ku between AD 58 and 76. This contains a wide variety of astronomical records, largely solar eclipses and comets, but also lunar and planetary events.

In the remainder of this chapter we shall restrict our attention to the astronomical records in three works – the *Ch'un-ch'iu*, *Shih-chi* and *Han-shu*. Because of the ambiguity of interpretation and uncertainty of date for any preserved observations before the *Ch'un-ch'iu* period, we shall not consider these further. Our period of interest thus extends from 722 BC to AD 9.

Records of astronomical phenomena

These may be conveniently divided into two groups, Pre-Han and Han.

Pre-Han observations (722 to 207 BC)

The data in the *Ch'un-ch'iu* is far superior to that found in subsequent centuries down to the beginning of the Han Dynasty. If this is a typical example of a state chronicle of the period, it is evident just how disastrous was the loss of other similar works both during the Burning of the Books in 213 BC and the subsequent destruction of the Ch'in capital in 206 BC.

The *Ch'un-ch'iu* lists notable astronomical events along with state affairs in strict chronological order. The earliest observation which it reports is that of a solar eclipse. This record is brief but it gives an exact date. We read: 'During the third year of Duke Yin of Lu in the spring, the second month on the day *chi-szu*, the Sun was eclipsed.' The date corresponds to 22 February 720 BC, and on this day, according to modern calculations, a partial solar eclipse was visible in Lu. In all, thirty-seven solar eclipses are reported in the chronicle, the last on 19 April 481 BC. Nearly all recorded dates agree exactly with the calculated dates of eclipses and this is an impressive series of observations for such an early period. It seems likely that the Lu astronomers kept a special watch for solar eclipses around the beginning of each lunar month, partly to check their calendar but also because of the astrological importance attached to these events. At least as early as the eighth century BC an eclipse of the sun – unlike a lunar eclipse – was regarded as an unfavourable omen. Thus the *Shu-ching* or 'Book of Odes' in describing an eclipse occurring around 750 BC remarks: 'For the Moon to be eclipsed is but an ordinary matter; now that the Sun has been eclipsed, how bad it is!'

The *Chu'un-ch'iu* records special ceremonies held at the Lu court at the time of the solar eclipses of 668, 664 and 611 BC. On each occasion, drums were beaten and sacrifices offered. A similar ceremony was performed in 668 BC when there were severe floods. Three of the eclipses – in 709, 601 and 549 BC – were described as total but unfortunately there are no further details.

Only three comets are reported in the *Ch'un-ch'iu* (in the years 613, 525 and 482 BC) and it thus seems that comets attracted much less attention than eclipses. The first of these, the earliest extant cometary observation from China which has any claim to reliability, was recorded as follows:

> 'During the fourteenth year of Duke Wen of Lu, in autumn, the seventh month, a bushy star entered Pei-tou [the Dipper].'

It will be noticed that only the month is mentioned, possibly because the comet remained visible for several days or weeks. However, the observers did name the asterism in which the comet appeared. The record in 525 BC also gives a fairly precise location 'in the vicinity of Ta-ch'en [Antares]', but this time only the season of the year, winter, is stated. Finally in 482 BC we have again the month of observation, the eleventh, but only the general direction in which the comet appeared, the east. All three comets were described as 'bushy stars' but this may well have been a general term for a comet at this early period. Unfortunately in no case do we have a possible reference to Halley's Comet, whose calculated dates of appearance around these times were 616, 540 and 466 BC. The only other astronomical records in the *Ch'un-ch'iu* describe a meteor shower in 686 BC when 'stars fell like rain' and a fall of meteorites, five in number, in 644 BC.

Most of the astronomical records in the *Shih-chi* are contained in the *Piao* ('Chronological Tables'). These are given in summary form along with non-astronomical events. However, additional details are occasionally included in the *Pen-chi* ('Basic Annals') and *Lieh-ch'uan* ('Biographies'). The *Shih-chi* cites nine solar eclipse observations between 443 and 247 BC, but only in two instances do we find even the month of occurrence noted. However, it is interesting that in 443, 382 and 301 BC the texts state that on account of the eclipse it became dark in the daytime, stars becoming visible on the first occasion. Fortunately it is possible to calculate the exact dates on which a large eclipse occurred even if only the year is given, since these phenomena are fairly rare in any one area. Hence the three observations can be accurately dated to 24 October 444 and 3 July 382 and 26 July 300 BC.

If even the dates of eclipses are reported so inexactly, we may not expect anything better for comets, and such is the case in practice. Between 516 and 214 BC the *Shih-chi* notes a total of twelve comets. However in only two cases, 240 and 234 BC, is the month specified while in a third, 482 BC, the season, winter, is noted. In an isolated instance, 238 BC, the asterism in which a comet was seen is named, Nan-tou, in Sagittarius, but other reports at best mention only the general direction in the sky. It is thus obvious how difficult it is to trace appearances of Halley's Comet in such unpromising material. The comets of 467 and 240 BC, reported only in the *Shih-chi*, have been regarded as possible sightings of Halley's Comet. However, as we shall see, only on the latter date is there any real justification for such a claim.

Few astronomical phenomena apart from solar eclipses and comets are noted in the *Shih-chi*, but two of these are so intriguing that they are worth quoting in full. They give a fascinating insight into the attitude to celestial events which was prevalent at the time.

The biography of the rulers of Sung, one of the Warring States, remarks that in the thirty-seventh year of the reign of Duke Ching (480 BC) Mars guarded Hsin (three stars in Scorpius including Antares). The text continues:

> 'Hsin is the celestial division representing the state of Sung. Duke Ching was worried. The astronomer Tzu-wei said to him, "We can move [the calamity] onto the ministers." Duke Ching replied, "The ministers are essential parts of me (literally, my legs and arms)." [Tzu-wei] said, "We can move [the calamity] onto the people." Duke Ching replied, "The ruler must take care of his people." [Tzu-wei] said, "We can move [the calamity] onto the harvest." Duke Ching replied, "If there is a famine, the people will be in distress. Who am I to be a ruler to?" Tzu-wei said, "Heaven is high but it listens to those below. Your Highness has given three wise sayings of a princely man, Mars must change its position." Later, when they observed it, it had indeed moved three degrees.'

Much later, in 211 BC the Annals of Ch'in Shih-huang record the following:

> 'In his thirty-sixth year, Mars guarded Hsin. A falling star came down in the eastern provinces. When it reached the ground it became a stone.

Someone engraved on the stone these words: "When Ch'in Shih-huang dies, the land will be divided." On hearing about this, Ch'in Shih-huang dispatched an official to investigate the matter. No one would answer so he arrested the people living in the neighbourhood of the stone and put them to death. Then he destroyed the stone by fire. . . .'

Less than a year afterwards the Emperor died while on a tour of the eastern provinces and within five years the Ch'in Dynasty was brought to an end. Ch'in Shih-huang's tomb may still contain a star map and other relics of scientific interest. The description in the *Shih-chi* makes fascinating reading:

'Ch'in Shih-huang ordered the artisans to make mechanised bows and arrows [to guard the entrance], so that if anyone came near they would automatically fire. Mercury was used to simulate the flow of the many rivers, [Yang-Tze] Kiang, [Huang] Ho and the great sea. Machines were used to circulate [the mercury] and make it flow. Above (on the ceiling) astronomical charts were drawn. Below (on the floor) geographical maps were depicted.'

The Former Han Dynasty (202 BC to AD 9)

Although this is the earliest period in Chinese history from which a wide variety of observations has survived, Han astronomical records are often deficient over long periods. Detailed accounts of celestial phenomena are, in fact, quite rare. Among recent archaeological finds is a silk manuscript found in a Han tomb which gives fairly extensive observations of the heliacal rising and setting of the various planets over the period from 246 to 177 BC. However, nothing as detailed as this occurs in the *Han-shu* itself. No doubt by the time the history was composed material such as this had already been lost. In addition, the content of a history is dependent on what its author thought fit to preserve and it is possible that much material which might have been of considerable astronomical importance at the present day was deliberately omitted.

The main source of observations from this period is the official history, the *Han-shu*. This contains a detailed 'Treatise on Astronomy', the *T'ien-wen Chih*, in which the bulk of the data is collected. However, solar eclipses are listed in a separate 'Treatise on the Five Elements', the *Wu-hsing Chih*, which also gives a few cometary reports not found in the astronomical treatise. Some of the observations in these two sections are duplicated in the *Pen-chi* or 'Basic Annals' at the beginning of the *Han-shu*. However, the annals also contain much additional material not found in the rest of the history, in particular a likely account of Halley's Comet in 87 BC. Unfortunately technical details are usually omitted in the annals, as is true of the *Lieh-chuan* or biographical section, which contains a few scattered astronomical records.

The five elements which are the subject of the *Wu-hsing Chih* are earth, wood, fire, metal and water. Thus wood and metal replace air in the list of four elements recognised by the ancient Greek philosophers. Each of the five bright planets was named by the Chinese after an element. Saturn was 'Earth Star', Jupiter 'Wood Star', Mars 'Fire Star', Venus 'Metal Star' and Mercury 'Water Star'. This set of names for the planets first appeared in the *Shih-chi*.

The main motive for making astronomical observations in the Pre-Han period was almost certainly astrological, but this theme is not usually obvious from the brief records themselves. However, Han observations reveal a deep concern with astrological matters. This was a time of growing acceptance of Confucian ideas that Heaven constantly watches over state affairs and consequently sends warnings whenever any wrong occurs. These warnings could either be celestial (any of the kind noted above) or terrestrial (e.g. earthquakes, droughts, locust plagues, snowfall in summer). Of the celestial phenomena noted only eclipses had any real practical importance – i.e. in regulating the calendar – so that without this widespread belief in portents little else might have been noted. Possibly the one redeeming feature of the pseudo-science of astrology is that in various parts of the world it has been responsible for the observing and recording of many celestial phenomena which probably would otherwise have passed unnoticed.

It was not uncommon for an emperor to issue an edict following the occurrence of a comet or other major astronomical phenomenon. A typical example in the *Han-shu* annals dates from 44 BC: 'In the summer, the fourth month [of the fifth year of the Ch'u-yuan reign period], a bushy star appeared in Shen (Orion) and an imperial edict said, "Due to our inadequacy the officials are not properly inspected. There are many vacant offices on account of the absence of suitable persons. The common people are disappointed. This is felt by the High Heaven above, such that the Yin and Yang have shown changes. Our fault has spread to all the people. We are greatly dismayed . . .".'

Presumably edicts such as these were drafted by Confucian ministers who were already aware of misgovernment and hence took the opportunity to implement changes in policy. Obviously such situations could easily be manipulated by corrupt officials and this might lead us to question some of the astronomical records. However, apart from solar eclipses the record of astronomical phenomena in the *Han-shu* seems reliable enough. Thus almost all of the lunar and planetary observations reported can be verified by modern calculation. It does seem possible that at various periods a number of unfavourable omens may have been passed by in silence, but there is little to suggest deliberate fabrication of observations. On the other hand Han solar eclipse records tend to be unreliable; as many as one quarter cannot be identified by calculation. However, we do know that attempts were made to predict eclipses using numerical cycles. In later dynasties such prediction became reasonably accurate, but in Han times it was largely unsuccessful. Hence it is possible that most of the false records represent no more than abortive predictions. We might imagine that once a prediction was made the astronomers felt obliged to report the occurrence of an eclipse in case they were accused of negligence. As far as we know, phenomena other than eclipses were not predicted, which may explain why we find a much more reliable record of lunar and planetary events.

Although many Han records of solar eclipses are untrustworthy, this is not true of those cases where a large fraction of the sun was obscured. Between 198 BC and AD 2, the *Treatise on the Five Elements* reports as many as ten eclipses in which the phase was total or very large. All of these can be readily identified by calculation. A common expression to describe a large partial eclipse is 'not complete, like a hook' – a rather apt expression. Only in a single instance is there any dating error. One of the total eclipses (181 BC) is said by the *Shih-chi* to have caused darkness in the daytime. Both this work and the *Treatise* remark that the unpopular Empress Dowager showed aversion from it, saying 'This is for me'. The following year she died and the omen was thus regarded as being fulfilled.

Other interesting observations made by the Han astronomers include a number of detailed descriptions of the motion of various comets through the constellations, notably in 157, 147, 138, 49 and 12 BC. Several careful accounts of planetary movements are also given, for example, apparent close approaches of Mars to Venus (147 BC) and Jupiter (29 BC). In the former case the estimated separation was only 1 *ts'un* (roughly 0.1°) while in the latter the two planets were only half this distance apart. Occultations of Mars and Saturn by the moon are reported in 69 and 29 BC respectively while in 28 BC we find what is probably the earliest reliable record of a sunspot from anywhere in the world. This is described as follows:

'In the third month [of the first year of the Ho-p'ing reign period], on the day *i-wei*, the Sun appeared yellow. There was a black vapour like a coin in the middle of the Sun.'

Dates of solar eclipses are always specified to the very day itself but for other phenomena – including comets – it is quite common to give only the month of occurrence. There is a particularly unsatisfactory period between 132 and 78 BC during which it is usual to indicate only the reign period in which an event occurred. Judging from this evidence alone, it is apparent just how uneven were the astronomical data which were available when the *Han-shu* was compiled.

We have discussed Han astronomical records in some detail but so far have made no mention of the astronomers themselves. These were civil servants whose chief task was to maintain a regular watch of the day and night sky for any celestial omens which might occur. Each phenomenon has its own special interpretation and whenever a portent was noted, it was necessary to warn the emperor as soon as possible so that the appropriate action might be taken. The office of court astronomer was not usually hereditary and in order to achieve such a position it would be necessary to pass the appropriate examinations. In Han times despite the great importance attached to their observations the astronomers were graded fairly low among officials. However, it is clear from some of the observational records that they had been fairly well trained.

It is all the more regrettable that what must be only a minute proportion of the records of the imperial astronomers has survived. Thus while we have nightly reports in the astronomical diaries from Babylon, no more than three observations in any one year are extant from China at any time during the whole of the Former Han Dynasty. In later periods Chinese observations become much more frequent, but this is certainly not the case in the more ancient times which paralleled the Babylonian texts. Quite probably, official astronomers were employed by the rulers of the various Warring States in Pre-Han times, but since so much information has been lost from this early period we know very little about them or their activities.

The Chinese calendar

Like many ancient calendars, the Chinese calendar was luni-solar. Years were not numbered continuously from some standard epoch; instead, they were counted from the beginning of each individual reign or reign period. Sub-division of reigns first began early in the Former Han and since this time there have been several hundred reign periods, some lasting less than a year. A number of chronological tables are readily available. These give the equivalent BC and AD dates with high accuracy back to the beginning of the Ch'un-ch'iu period.

For most of the period under discussion the year began with the first month, which corresponds to January or February in the Western calendar. However, in the calendar of Ch'in Shih-huang, the year began with the tenth month and this continued in use in the early part of the Han Dynasty (221 to 104 BC). Individual years were either of twelve or thirteen months' duration. As a common year of twelve lunar months contained about 354 days, in order to keep the calendar roughly in step with the seasons an extra month was intercalated roughly every thirty months or so. Intercalation in much of the period before the Han was somewhat irregular, but during the Han a precise scheme was adopted, making accurate date conversion a relatively simple matter.

Lunar months contained either twenty-nine or thirty days, the average length being close to the mean synodic month of 29.53 days. However, in expressing individual dates the day of the month was not used unless it happened to be the first or last day. From a very early period – well before 1000 BC – an independent sixty-day cycle having close parallels with our much shorter week was employed.

A complete date was thus normally specified in terms of: (i) year of reign or reign period, (ii) lunar month and (iii) cyclical day. The use of cyclical days considerably simplifies conversion to the Western calendar, especially in the earlier period when intercalation was irregular. It is therefore a pity that so few early cometary records give more than the month of occurrence.

The Chinese view of the constellations

The Chinese representation of the constellations was radically different from that in the West, as a glance at any oriental star map reveals. Of the well-known star groups only the Plough (*Pei-tou*, the Northern Dipper) and Orion (*Shen*, the White Tiger) are recognisable. No really ancient star maps of the Pre-Han era now survive, but a variety of Far Eastern charts have been preserved from later centuries. As a result of careful measurements on these it is possible to identify several hundred of the brighter stars with a high degree of reliability. One of the most remarkable star maps is the Soochow Astronomical Chart. This was engraved on stone in AD 1247 for the instruction of the Chinese emperors. Possibly one of the best preserved maps in existence is the Tun-huang star map of the T'ang Dynasty (now in the British Library).

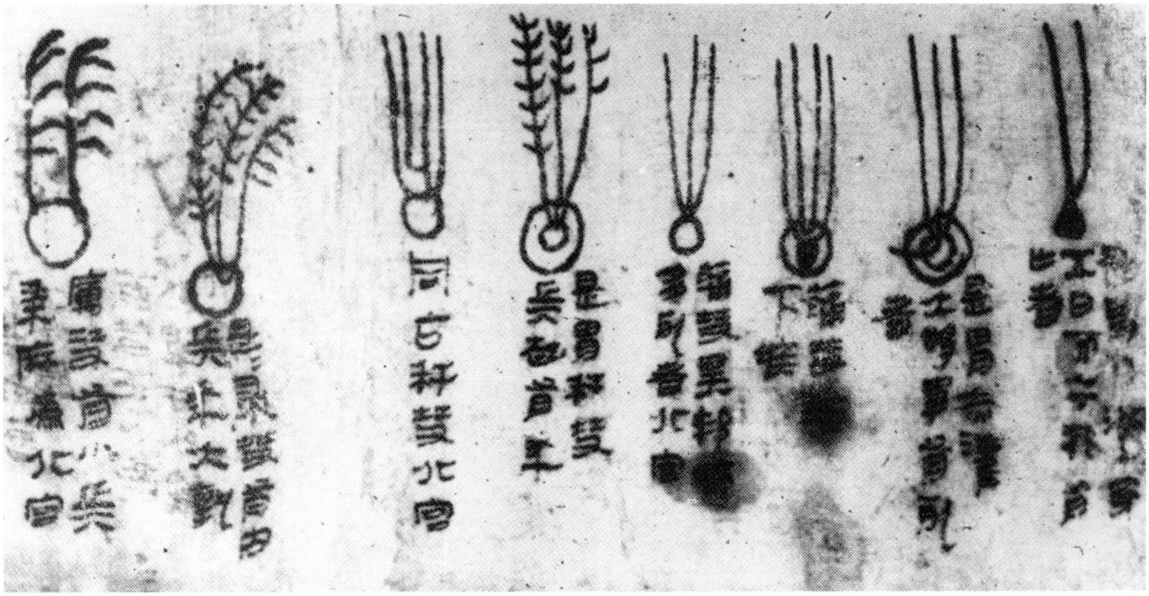

Chinese drawings of the various forms which a comet can take. From a tomb of the Former Han Dynasty (date of burial 168 BC)

In order to identify stars and facilitate observation and recording of astronomical phenomena, stars were grouped together in somewhat irregular numbers to form asterisms, known as *Hsing-kuan* by the Chinese. The number of stars to an asterism could vary from only one to usually not more than ten. There were several hundred asterisms, and within certain prominent asterisms each individual star had an established name. In addition near, but not identical with, the zodiacal belt, twenty-eight unequal zones along the ecliptic called 'lunar mansions' were used to define co-ordinate zones for measurement purposes. Within each lunar mansion there dwelled several *Hsing-kuan* (star officials) – asterisms. In a historical record mention of a few of the asterisms near which a comet passed, together with the appropriate dates, may allow its motion to be studied in some detail.

Records of Halley's Comet

Between 613 BC – the earliest cometary record in the *Ch'un-ch'iu* – and the end of the Former Han Dynasty, more than fifty comets are recorded in Chinese historical works. These are mainly described as *hui-hsing* ('broom star') or *po-hsing* ('bushy star'). In later times, the term broom star was used to denote a comet with a developed tail, while a bushy star had no obvious tail. However, at this early period it is doubtful whether these terms were carefully distinguished. In modern times bright comets have been seen to occur fairly frequently and only a small proportion of Chinese records can possibly relate to Halley's Comet. It is by no means an easy task to distinguish actual sightings of Halley's Comet among the more ancient cometary records.

The calculated dates of return of Halley's Comet between 722 BC and AD 9 according to Donald Yeomans and Tao Kiang are as follows: 690, 616, 540, 466, 391, 315, 240, 164, 87 and 12 BC. Within a year or two of these dates there are Chinese cometary sightings only in 467, 240, 162, 87 and 12 BC. As indicated previously the 12 BC observation is very reliable, but the earlier records listed here are comparatively vague. We shall consider each of the selected dates in turn.

467 BC

This comet is recorded only in the Chronological Tables of the *Shih-chi* and reads as follows:

'During the tenth year of Ch'in Li-kung, a broom star was seen.'

The chronology of this period is very secure but the record is so brief that even if the calculated year of appearance of Halley's Comet agreed exactly with the recorded year, it would be impossible to prove that this comet was indeed referred to. It could well have been some unknown long period comet. For comparison, only about three months before the long awaited return of Halley's Comet in 1910 a brilliant daylight comet appeared. Again, in AD 760 the astronomers of China were disturbed to find two separate comets visible at one and the same time. As it happens, Halley's Comet should not have been visible to the unaided eye during 467 BC, the calculated date of perihelion being July of the following year. There seem to be too many problems connected with this record to warrant further consideration.

240 BC

This, the very earliest plausible reference to Halley's Comet, is still rather brief. The Annals of the *Shih-chi* note that in the seventh year of Emperor Ch'in Shih-huang:

> 'A broom star first appeared at the eastern direction; it was then seen at the northern direction. During the fifth month (May 24 – June 23) it was again seen at the western direction. General Ao died [while at war] . . . The broom star remained visible at the western direction for sixteen days. In the summer, the Empress Dowager died.'

A summary of this account containing no extra information is given in the Chronological Tables of the *Shih-chi*.

Calculation shows that Halley's Comet would probably become visible in mid-May in 240 BC and would be seen before dawn in the eastern sky. By the end of the month when it made its closest approach to the Earth, it might have been seen in a roughly northern direction (north-north-east) before dawn. Following conjunction with the sun at the beginning of June, the comet would be seen in the western sky after dusk and would be particularly well placed for observation in mid-June. The reported duration of sixteen days at the west is quite in keeping with calculation; towards the end of June, Halley's Comet would have faded considerably. The agreement between observation and the expected motion on this basis is impressive and we can be fairly confident that this is truly a sighting of Halley's Comet.

162 BC

As shown in chapter 2, Halley's Comet was carefully observed in Babylon in the autumn of 164 BC. However, the *Han-shu* does not mention any comet appearing during this year or the immediately preceding and following years. The Astronomical Treatise of the *Han-shu* mentions the sighting of what may have been a comet on a date corresponding to February 6 in 162 BC, but this is much too late for Halley's Comet. This reads as follows:

> 'In the second year of the Hou reign period of Han Wen-ti, first month, day *jen-yin* (February 6) a *T'ien-ch'an* (comet) appeared at the south-west.'

According to modern calculations, Halley's Comet should have appeared in the autumn of 164 BC, and the reason why we have no Chinese account of this return may well be that practically all observations from around this time were lost. Between 192 and 157 BC only three astronomical records are preserved.

87 BC

There is a short record of a comet in the *Han-shu* annals under the second year of the Hou-yuan reign period (87 BC). It reads as follows:

> 'In autumn during the seventh month (August 10 to September 8) there was a bushy star at the east.'

No parallel entry is found in the Astronomical Treatise of the *Han-shu*. This is again a time of few astronomical records, only six being noted between 132 and 78 BC.

If this record indeed refers to Halley's Comet then either the month or the direction must be in error. Halley's Comet should have been visible in the east for most of July, reaching conjunction with the sun before the end of the month. By the beginning of August it would have reappeared in the western sky after dusk. An error in the month seems to be the most likely possibility. Reading sixth month for seventh (although the characters for six and seven are dissimilar) would put the observation at some time between July 12 and August 9. This would fit in well with the calculated dates of visibility in the east and also the Babylonian observation.

12 BC

After the disappointing observations at previous apparitions, it is encouraging to come across such a detailed account of the motion of Halley's Comet, which was followed for as long as fifty-six days. The account in the Treatise on the Five Elements of the *Han-shu* reads as follows:

> 'In the first year of the Yuan-yen reign period, the seventh month, on the day *hsin-wei* (August 26) there was a bushy star at Tung-ching; it was treading on Wu-chu-hou. It appeared to the north of Ho-shu and passed through Hsuan-yuan and T'ai-wei. Later it travelled at more than 6° daily. In the morning it appeared at the eastern direction. On the evening of the thirteenth day (September 7) it was seen at the western direction. It trespassed against Tz'u-fei, Ch'ang-ch'iu, [Pei-] Tou and Saturn. Its "swarming flames" again penetrated within Tzu-wei, with Ta-huo (Antares) right behind. It reached T'ien-ho (the Milky Way), sweeping the region of Hou and Fei. It moved south, crossing and trespassing against Ta-chueh (Arcturus) and She-t'i. When it reached T'ien-shih it moved slowly at a regular pace. Its "flames" entered T'ien-shih. After a further ten days it

七年。彗星先出東方。見北方。五月見西方。〔正義〕彗音似歲反見音行練反……經內記云彗出北斗兵大起……台臣害君彗在太微君害臣彗在天獄諸侯作亂所指其處大惡彗在口旁子欲殺父

將軍驁死。以攻龍、孤、慶都。〔集解〕徐廣曰……一作應〔正義〕括地志云定州恒陽縣西南四十里有白龍水又有挾龍山又定州唐縣東北五十四里有孤山蓋都山也帝王紀云望堯母慶都所居張晏云堯山在北堯母慶都山在南相去五十里北登堯山南望慶都山也注水經云望都故城東有山不連陵名之曰孤孤都聲相近疑即都山孤山及望都故城三處相近。

還兵攻汲。彗星復見西方十六日。夏太后死。〔索隱〕莊襄王所生母也。〔正義〕子楚母也。

Shih-chi, Annals; description of Halley's Comet in 240 BC

went towards the west. On the fifty-sixth day (October 20) it went out of sight together with Ts'ang-lung.'

An imperial edict following the apparition of the comet gives a splendid insight into the official attitude of the time. We read:

'Lately, reproaches in the form of solar eclipses and meteors have been seen in the sky. These great strange signs were repeated and yet those in official positions remained silent; rarely has there been loyal advice. Now a bushy star has been seen in Tung-ching. We are very dismayed. The ministers, grandees, doctors and advisors are each to think solemnly as to the meaning of these changes and compare them clearly with the Classical texts; nothing is to be concealed. These and the inner prefectures are each to recommend one person of good moral character who can remonstrate without reservation. Let the twenty-two northern prefectures each recommend one person who is brave and fierce and an expert in military tactics.'

Apart from one or two minor details the observed path of the comet through the constellations from Gemini to Scorpius agrees well with the calculated motion of Halley's Comet. From this time onwards, most Chinese accounts of the comet give exact dates, usually with accurate positional information, so that the identity of the comet is seldom in doubt.

Ch'ien-han-shu, Treatise on the Five Elements; description of Halley's Comet in 12 BC

[illegible]灘南苦反隷八月長星出于東方長終天三十日去占曰是為蚩尤旗見則王者征伐四方其後兵誅
四夷連數十年元狩四年四月長星又出西北是時伐胡尤甚元封元年五月有星孛于東井又孛于三
台其後江充作亂京師紛然此明東井三台為秦地效也宣帝地節元年正月有星孛于西方去太白二
丈所劉向以為太白為大將彗孛加之掃滅象也明年大將軍霍光薨後二年家夷滅成帝建始元年正
月有星孛于營室青白色長六七丈廣尺餘劉向谷永以為營室為後宮懷姙之象彗星加之將有害懷
姙絕繼嗣者一曰後宮將受害也其後許皇后坐祝詛後宮懷姙者廢趙皇后立妹為昭儀害兩皇子上
遂無嗣趙后姊妹卒皆伏辜元延元年七月辛未有星孛于東井踐五諸侯孟康曰五諸侯星名出河戍北率行軒
轅太微後日六度有餘晨出東方十三日夕見西方犯次妃長秋斗填蠭炎再貫紫宮中大火當後達天
河除於妃后之域南逝度犯大角攝提至天市而按節徐行服虔曰謂行遲炎入市中旬而後西去五十六日與
蒼龍俱伏谷永對曰上古以來大亂之極所希有也察其馳騁驟步芒炎或長或短所歷奸犯師古曰奸音干內
為後宮女妾之害外為諸夏叛逆之禍劉向亦曰三代之亡攝提易方秦項之滅星孛大角是歲趙昭儀
害兩皇子後五年成帝崩昭儀自殺哀帝卽位趙氏皆免官爵徙遼西哀帝亡嗣平帝卽位王莽用事追
廢成帝趙皇后哀帝傅皇后皆自殺外家丁傅皆免官爵徙合浦歸故郡平帝亡嗣莽遂篡國盩公十六

CHAPTER 4

An outline history of Halley's Comet prior to the telescopic period

In this chapter we discuss briefly some of the more important observations of Halley's Comet at each return from 240 BC to AD 1682. The comet was most extensively watched in China during this period and we shall thus place a special emphasis on Far Eastern sightings. For the sake of completeness we summarise our discussion in chapters 2 and 3 of the four earliest returns. All calculations are based on the orbital details computed by Donald K. Yeomans and Tao Kiang, which seem to fit both ancient and medieval observations extremely well. At the beginning of each entry the calculated date of perihelion and the nearest approach to the Earth are listed. For comparison the average distance between the sun and Earth is about 150 million km.

240 BC
Date of perihelion: May 25
Least distance from Earth: 67 million km on June 3
This is the earliest identifiable return of Halley's Comet. The only surviving observation is a brief Chinese record. This states that the comet first appeared in the east at some time between May 24 and June 23. Later after conjunction with the sun it was observed in the west for a total of sixteen days. It is probable that the Babylonian astronomers observed the comet carefully at this and even previous returns. Unfortunately as few of the earlier Babylonian astronomical diaries have yet come to light, there are no known Babylonian cometary records before 234 BC. The brief Chinese observation agrees well with modern calculation.

164 BC
Date of perihelion: November 12
Least distance from Earth: 16 million km on September 28
On this rather close approach to the Earth Chinese records are silent, probably on account of the loss of most of the astronomical records from around this time, long before the official history of the period was written (first century AD). Fortunately two separate Babylonian tablets provide overlapping descriptions. These state that the comet first appeared in the east in the Pleiades area of Taurus and later was seen in the west in Sagittarius, passing a few degrees to the north of Jupiter. The date of first sighting is not given but we are told that the comet reached Sagittarius at some time between October 21 and November 19.

87 BC
Date of perihelion: August 6
Least distance from Earth: 66 million km on July 26

Here for the first time we encounter two independent observations of Halley's Comet from widely separated parts of the world. The Chinese record is poor and probably contains a dating error. It merely states that a comet was seen at the east. The Babylonian record was probably quite detailed but only a fragment of it has survived. From this we learn that for some time between July 14 and August 11 the comet had moved at a fairly regular pace and that its tail had reached a length of about 10°. It was probably last seen in the evening sky on August 24.

12 BC

Date of perihelion: October 10

Least distance from Earth: 24 million km on September 9

There are no surviving Babylonian diaries after about 40 BC and although the astronomers of Babylon may have studied the comet carefully at this fairly close return nothing of their records now remains. However, the Chinese astronomers have furnished us with the earliest detailed account of the motion of the comet, from Gemini to Scorpius. The exact dates of first and last visibility (August 26 and October 20) are given, a total of fifty-six days duration. In his *Roman History*, Dion Cassius makes the earliest known mention of the comet from Europe. Among a number of portents preceding the death of Agrippa, 'The star called comet hung for several days over the city and was finally dissolved in flashes resembling torches.'

AD 66

Date of perihelion: January 26

Least distance from Earth: 37 million km on March 19

Once again, the only careful description of this return is from China. The Chinese description is rather fragmentary, but from it we learn that it was first sighted on January 31, later appeared in Capricorn, and finally disappeared around April 10, having passed through several zodiacal signs. The Jewish historian Josephus, who lived through the siege and fall of Jerusalem (AD 70), mentions several portents in years leading up to these events. Among these, 'a star, resembling a sword, stood over the city'. It is possible that this was an allusion to Halley's Comet in AD 66, but this can be no more than speculation.

AD 141

Date of perihelion: March 22

Least distance from Earth: 25 million km on April 22

The only known record of this return is from China. However, this account gives several interesting details such as the colour of the comet – a 'pale blue'. We also find a single fairly accurate measurement of the position of the nucleus – correct to the nearest degree. The comet was first seen in the east on March 27. By April 22 it had passed conjunction with the sun and was now visible in the west at dusk. The comet was last seen in Leo, having moved fairly rapidly eastwards from the Pisces region.

AD 218

Date of perihelion: May 17

Least distance from Earth: 63 million km on May 30

No accurate dates are available for this return, the Chinese record giving only the month of first sighting (between April 14 and May 12). When first seen, it was visible in the western sky for more than twenty days. Later it became an evening object and a description of the motion between Auriga and Virgo is given. In Europe Dion Cassius, writing as a contemporary, lists a comet among a number of portents occurring shortly before the death of the Emperor Macrinus, who was killed on June 7: 'For a very distinct eclipse of the Sun occurred just before that time and a comet was seen for a considerable period. . . .' This statement is a little confused, for the only large eclipse visible in Rome over a period of several years occurred on October 7. Nevertheless it seems probable that the comet referred to was indeed Halley's, and here the timing is fairly good.

AD 295
Date of perihelion: April 20
Least distance from Earth: 48 million km on May 12
Once again only an approximate date is available. Chinese records assign only the month of discovery, corresponding to May 1–30, and there is no estimate of the duration of visibility. First visibility was in Andromeda and the comet was last seen in Leo.

AD 374
Date of perihelion: February 16
Least distance from Earth: 13 million km on April 1
Around the time of this very close approach to the Earth, Halley's Comet would be well placed for viewing, almost opposite to the sun. It must have been a spectacular sight, but the brief Chinese records which have come down to us do not convey this impression. They tell only of a bushy star appearing in Aquarius on a date corresponding to March 6, which was later identified as a broom star by the time that it had reached Libra on April 2. A second comet was seen in the same year by the Chinese astronomers – on November 19. This must have been a separate object from Halley's Comet, which by November would have long since faded out of sight and in any case was in a different part of the sky.

AD 451
Date of perihelion: June 28
Least distance from Earth: 73 million km on June 30
There are two independent accounts from China of this return, and also an unusually careful record from Europe. Combining the two reports from north and south China, the period of visibility was more than two months – from June 10, when the comet was in Taurus, to August 15, by which time it had reached Virgo. In Europe the comet was seen around the time of the Battle of Chalons, at which the Roman general Aetius defeated Attila the Hun. The record states that a comet began to appear on June 10. On June 29 it was observable both in the morning and evening, implying that conjunction with the sun occurred on this day. Finally the comet was lost to view on August 1.

AD 530
Date of perihelion: September 27
Least distance from Earth: 42 million km on September 3
Here the Chinese record is much more precise, but there are some interesting descriptions by Byzantine chroniclers. The comet was first seen in China on August 29 in the constellation of Ursa Major. Its colour was described as pure white. On September 1, the comet was observed only about 1° from Nu Ursae Majoris, a good positional record. By September 23 it had faded considerably and was lost to view four days later. Byzantine writers noted that the comet was first seen in September and remained visible for a total of twenty days. It was described as 'a huge and terrible star', whose rays extended towards the zenith. Because it resembled a burning torch, it was called 'Lampadias', a name previously used by Pliny to describe a comet of this form.

AD 607
Date of perihelion: March 15
Least distance from Earth: 13 million km on April 19
This unusually close approach does not seem to have been noticed in Europe and even the Chinese records are surprisingly vague. Judging from the latter, there would appear to have been two (or even three) comets visible in this year and it is by no means easy to disentangle the references to Halley's. Assuming plausible dating errors, the comet was first seen on March 30 in Pegasus, and after conjunction with the sun it was followed for a further period as far as Gemini. A detailed account is also given of a separate comet which was sighted on October 21 and remained visible for more than two months.

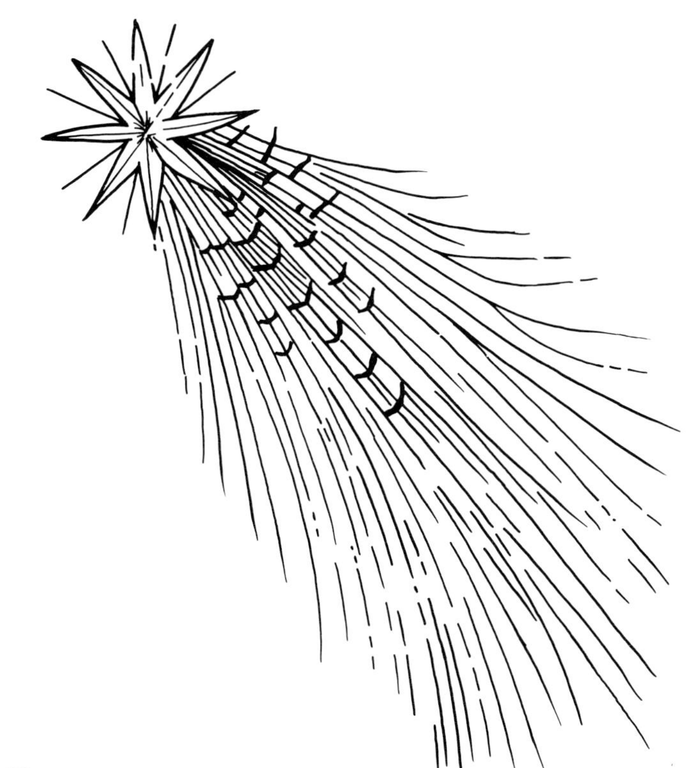

A comet engraved for the *Nuremburg Chronicle* by Michael Wohlegemut and Wilhelm Pleydenwurff in 1493. It appears on the page dealing with AD 684 and is repeated elsewhere in the book

AD 684
Date of perihelion: October 2
Least distance from Earth: 39 million km on September 7
The brief Chinese records do not mention a single asterism through which the comet passed. All that can be established is that a comet with a tail more than 10° in length was first sighted on September 6 in the western sky. By October 9 it had disappeared. For the first time we find a brief Japanese record, giving no further details. However this also notes a separate comet which appeared near the Pleiades in mid- or late December. By early January it had disappeared. This could not have been Halley's, which by then would be very faint and on the opposite part of the sky from the Pleiades. Interestingly enough, in the *Life of Pope Benedict II*, it is recorded that in this year between Christmas and Epiphany there appeared near the Pleiades 'an absolutely shadowy star, resembling the Moon when covered with clouds'. This is clearly the same object as was reported by the Japanese. It would thus appear that there is no actual record of Halley's Comet from Europe at this return.

AD 760

Date of perihelion: May 20
Least distance from Earth: 61 million km on June 3

The return in this year was closely watched by the Chinese astronomers, who followed the comet for more than fifty days, beginning with its first appearance in Aries on May 16. The comet was last seen in Virgo more than fifty days later. On May 18, a separate comet was sighted in the opposite part of the sky and the astronomers were much disturbed by the sight of two such objects visible at the same time. The astrological consequences were regarded as serious. Byzantine chroniclers record only Halley's Comet, which, 'like a glittering beam', appeared for six days in the eastern sky and was seen in the west for a further twenty-one days. This was followed by an eclipse of the sun on August 15. The eclipse, which is here correctly dated, was very large in eastern Europe.

AD 837

Date of perihelion: February 28
Least distance from Earth: 5 million km on April 11

In this year Halley's Comet made by far the closest known approach to the Earth, not much more than ten times the moon's distance. As a result the motion of the comet was strongly disturbed by the Earth's gravitational field. A remarkable description of this fly-by is given by the astronomers of China. First seen in Aquarius, on March 22, its position was carefully measured almost daily between April 5 and April 13. When nearest to the Earth, the comet was moving across the sky at roughly 2° per hour and its tail was observed to be more than 60° in length. The tail was distinctly seen to branch into two parts. Once again a further comet was seen, this time in the autumn long after Halley's Comet had been lost to view. At this return Halley's Comet was also observed in Japan, various parts of Europe and also Iraq. But these descriptions cannot be compared with the detailed accounts from China. Because of its interest we give a full translation of the principal account in the *Chiu-t'ang-shu*, Astronomical Treatise:

> 'On the night of March 22 a broom star appeared at the east direction. Its length was more than 7° and it was situated in the first degree of Wei (lunar mansion 12); it pointed west. On the night of March 24 it was south-west of Wei. The length of the broom was more than 7° and the brightness of its rays was becoming fierce. It again pointed west. On the night of March 29 the broom was 8° in Wei. On the night of April 5 it was 3.5° in Hsu (lunar mansion 11). On the night of April 6 the length of the broom was more than 10°. It moved west in a straight line, pointing slightly south. It was 1.5° in Hsu. On the night of April 7 its length was more than 20° and its width was 3°. It was 9° in Nu (lunar mansion 10). On the night of April 8 it increased in both length and width; it was 4° in Nu. On the night of April 9 its length was more than 50°. It branched into two tails, one pointing towards Ti (lunar mansion 3) and the other concealing Fang (lunar mansion 4). It was 10° in [Nan] Tou (lunar mansion 10). On the night of April 11 the length of the broom was 60°. The tail was without branches and it pointed north. It was 7° in K'ang (lunar mansion 2). The Emperor summoned the Astronomer Royal and asked him the reason for these star changes. . . . That same night the length of the broom was 50° and its width was 5°. It was moving towards the north-west and pointing east. On the night of April 13 the length of the broom was more than 80°. It was [still] moving north-west and pointing east; it was 14° in Chang (lunar mansion 26). . . . On the night of April 28 the broom was 3° in length. It appeared to the right of Hsuan-yuan (near Regulus) and was pointing east. It was 7° in Chang.'

申夜危之西南彗長七尺芒耀愈猛亦西指癸丑夜彗在危八度庚申夜在虛三度半辛酉夜彗長丈餘
直行稍南指在虛一度半壬戌夜彗長二丈其廣三尺在女九度癸亥夜彗愈長廣在女四度三月甲子
朔其夜彗長五丈岐分兩尾其一指氐其一掩月在斗十度丙寅夜彗長六丈尾無岐北指在危七度文
宗召司天監朱子容問星變之由子容曰彗主兵旱或破四夷古之占書也然天道懸遠唯陛下脩政以
抗之乃勅尚食今後每日御食料分為十日其夜彗長五丈闊五尺却西北行東指戊辰夜彗長八丈有
餘西北行東指在張十四度詔天下放繫囚撤樂減膳避正殿先是群臣拜章上徽號宜並停癸未夜彗
長三尺出軒轅之右東指在張七度六月河陽軍亂逐李詠是歲夏蝗大旱八月丁酉彗出虛危之間十

Chiu-T'ang-shu, Astronomical Treatise; description of Halley's Comet in AD 837

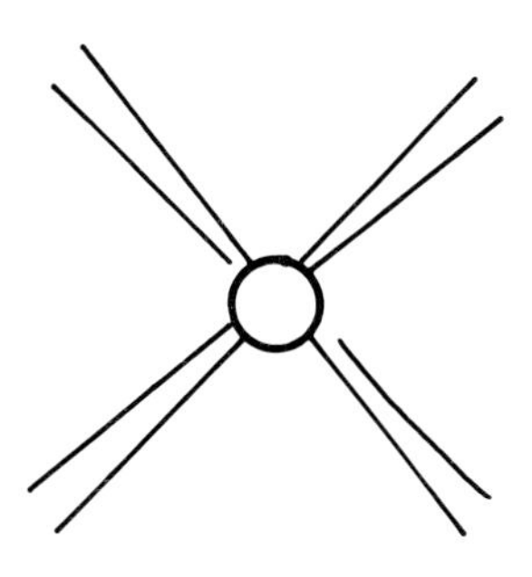

A comet with four tails, from Huang Ting's *Collected treatises on astronomy*

A briefer account in the *Hsin-t'ang-shu*, Astronomical Treatise, ends:

> 'On April 28 its length was 3°. It disappeared at the right of Hsuan-yuan. All broom stars appearing in the morning point towards the west and when appearing in the evening point towards the east; this is normal. However there had never been one pointing in all four directions and trespassing on so many [asterisms].'

AD 912

Date of perihelion: July 18
Least distance from Earth: 73 million km on July 16
Calculations suggest that the comet should have become visible during the second half of June. The only Chinese record of a comet in this year is in May. It appears that on this occasion the Chinese failed to record the comet. In Japan the comet was seen in the north-west from July 19 to 28, after conjunction with the sun, but its location in the sky is not recorded. The Baghdad chronicler Ibn al-Jawzi notes as many as three comets appearing in this year.

AD 989

Date of perihelion: September 5
Least distance from Earth: 58 million km on August 20
In China the comet was seen from August 12 to September 9, being observed in the east before conjunction and later to the north-west. The Chinese observations suggest that the date of perihelion may have to be corrected to September 9. According to an Arab chronicle (al-Maqrizi) 'a planet with a tail' was seen for twenty-two days from Cairo.

AD 1066

Date of perihelion: March 20
Least distance from Earth: 16 million km on April 24
The comet was first seen both in China and Japan on April 3, and the Chinese astronomers observed it both before and after conjunction for a total of sixty-seven days, which suggests that this was an unusually bright return. After perihelion they record it as having a broom-like vapour and resembling a one-tenth peck measure. A brief report in the *Koryo-sa* is the first reasonably reliable Korean observation of Halley's Comet; it describes it as being as large as the moon. It is recorded in local chronicles all over Europe and in Russia. In England it appeared before the Norman invasion, and is recorded on the Bayeux Tapestry observed by King Harold and his court; the caption reads *isti mirant stellā*, 'They wonder at the star.' According to the *Anglo-Saxon Chronicle*, 'Then there was seen all over England a sign such as no one had ever seen. Some said that the star was a comet, as some called the long-haired star, and it appeared first on the eve of the feast of the Greater Litany (April 24).' The cathedral archives at Viterbo in Italy record it as having a tail streaming like smoke up to nearly half the sky. At Baghdad its reappearance after conjunction is described by Ibn al-Jawzi, 'It re-appeared on Tuesday evening at sunset with its light folded around it like the moon. People were terrified and distressed.'

AD 1145

Date of perihelion: April 18
Least distance from Earth: 40 million km on May 12
In Europe the comet was seen first on April 15. In China it was sighted on April 26. The Chinese astronomers again observed the comet for a remarkable length of time after perihelion, until July 6. They describe it as coloured pale blue. The most detailed account of the comet on this occasion comes from Japan. A

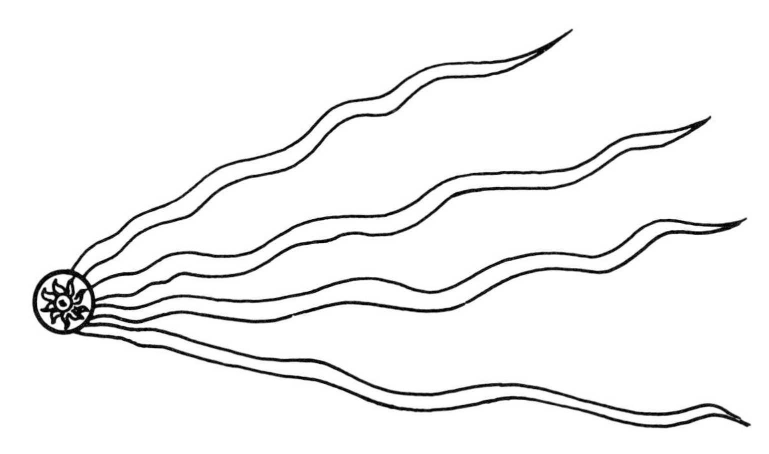

The comet drawn by Eadwine in the *Canterbury Psalter*, perhaps Halley's Comet in AD 1145

young courtier Fujiwara no Yorinaga was advised of the appearance of a new star by the professor of astronomy on May 3, and recorded his observations until June 18 in his personal diary, the *Taiki*.

A possible representation of the comet appears in the *Eadwine* (Canterbury) *Psalter*, now preserved at Trinity College, Cambridge. This illuminated manuscript, dated to the twelfth century, has a drawing of a comet at the bottom of the page on which Psalm 5 begins, and it is thought that the monk Eadwine included the picture because the comet appeared at the time he was completing the page. The Anglo-Saxon comment beside the drawing has been translated, 'Concerning the star called comet. A suchlike ray has the star known as Comet, and in English it is called "the hairy star". It appears seldom, after [periods of] many winters, and then for an omen.'

AD 1222

Date of perihelion: September 28

Least distance from Earth: 46 million km on September 6

The comet may already have been seen by European observers in August, but in China it was not seen until after conjunction with the sun on September 5. It certainly remained visible until October 8, but a report from south China that a comet appeared on September 25 and disappeared on October 23 cannot refer to Halley's Comet in view of its distance from the Earth in late October. In Korea it is said to have been seen by day, but this too seems unlikely; it would be the only recorded case of a daylight sighting although there have been much more favourable approaches of the comet in the last 2000 years.

From Mosul, Iraq, Ibn al-Athir records a large comet visible at dawn for ten days from September 29, reappearing in the west in the evening and remaining visible until early November. Taken with the report from south China this suggests that there were two comets visible this autumn.

AD 1301

Date of perihelion: October 25

Least distance from Earth: 27 million km on September 23

In Europe this appearance of the comet was described in Greek verse by the Byzantine scholar Georgius Pachymeres (*Andronicus Palaeologus* 4, 14); his poem is known to astronomers from a Latin translation by Petrus Possinus included in Pingré's *Cométographie* (1784). The Florentine chronicler Giovanni Villani (*Cronica universale* 8, 48) describes it as having 'great trails of smoke behind'. For scientific details we are again primarily dependent on the Chinese astronomers. They give an unusually precise measurement of the position of its nucleus on

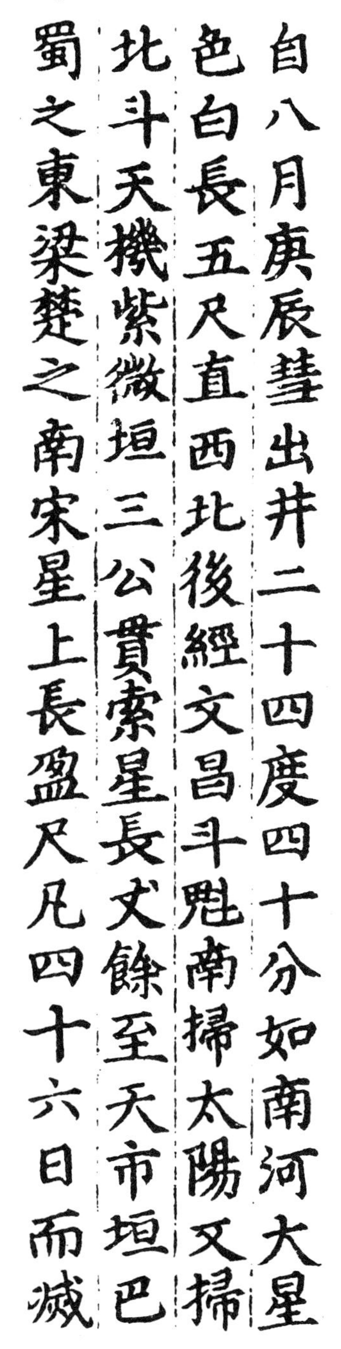
自八月庚辰彗出井二十四度四十分如南河大星
色白長五尺直西北後經文昌斗魁南掃太陽又掃
北斗天機紫微垣三公貫索星長丈餘至天市垣巴
蜀之東梁楚之南宋星上長盈尺凡四十六日而滅

Yuan-shih, Astronomical Treatise; description of Halley's Comet in AD 1301

September 16, shortly after discovery, 'A broom star appeared at 24.4° within Tung-Ching. It was like the large star of Nan-ho (Procyon). Its colour was white and its length was 5°.' It remained visible until October 31.

This return of the comet is now famous for having been depicted by the Florentine painter Giotto di Bondone in his painting *The Adoration of the Magi* in the Scrovegni (Arena) Chapel at Padua, Italy. Painted in 1303 it shows the Star of Bethlehem as a comet. Giotto in turn is commemorated by the European satellite mission to the comet in 1985–6.

AD 1378

Date of perihelion: November 10
Least distance from Earth: 18 million km on October 3
This is the last return of the comet for which the Far Eastern records are at all useful for study of the comet's motion. The Chinese saw the comet first on September 26. The Korean account comments on its close approach to the north celestial pole; its calculated path passed within 10° of the pole, nearer than on any other occasion in the last 2000 years. In Japan official prayers were offered on account of the comet on November 15 when it was presumably still visible.

AD 1456

Date of perihelion: June 9
Least distance from Earth: 67 million km on June 19
The Chinese saw the comet first on May 27. On this return Peurbach in Vienna made the first attempts to measure the comet's parallax (the varying angles at which it was seen from different locations). The most detailed observations were made at Florence by Paolo del Pozzo Toscanelli from June 8 to July 8, but these were not available to Halley; they were rediscovered in a library in Florence in 1864 and allowed the Italian astronomer Giovanni Celoria to establish the comet's orbit on this return with great precision. Toscanelli was the first astronomer to make a chart of the comet's movements. Contemporary records agree that on this occasion the comet had a remarkably long tail of some 60°. Its appearance is said to have alarmed the Turkish army of Mehmet II besieging Belgrade.

AD 1531

Date of perihelion: August 26
Least distance from Earth: 66 million km on August 14
The most accurate observations come from Peter Apian, astronomer to the Emperors Charles V and Ferdinand I of Austria, at Ingolstadt in Bavaria from August 13 to 23. He recorded the comet's altitude and azimuth each night at the time when the star Arcturus was at the meridian, and made a chart of the comet's daily movements. His observations were used by Halley to demonstrate that the comet which he saw in 1682 had also been seen in this year. Detailed records were also made by Paracelsus in Switzerland and by Girolamo Fracastoro at Verona. Both Apian and Fracastoro remarked that a comet's tail always points away from the sun. This had already been known to the Chinese by the seventh century AD, and maybe even earlier; their explanation was that comets shine by reflected light like the moon. Whereas the Chinese only remarked that the tails of comets seen in the eastern sky point to the west, and vice versa, Apian makes the additional point that the axis of the comet's tail if extended runs through the sun.

In 1577 another great comet appeared, and was studied at length by the Danish astronomer Tycho Brahe from the island of Hveen near Copenhagen. His measurements of the comet's parallax and the comparison of his observations with those made by Hagecius at Prague allowed him to provide the first scientific proof that comets moved outside the immediate sphere of the Earth and the moon. He concluded that the comet's orbit lay between the moon and Venus. For the first time the study of comets could be pursued on a realistic basis.

AD 1607

Date of perihelion: October 27
Least distance from Earth: 36 million km on September 29
For his calculation of the orbit of the comet on this occasion Halley used the observations made by Johannes Kepler at Prague from September 26 to October 26, and by Christian Longomontanus at Malmo, Sweden, and Copenhagen from

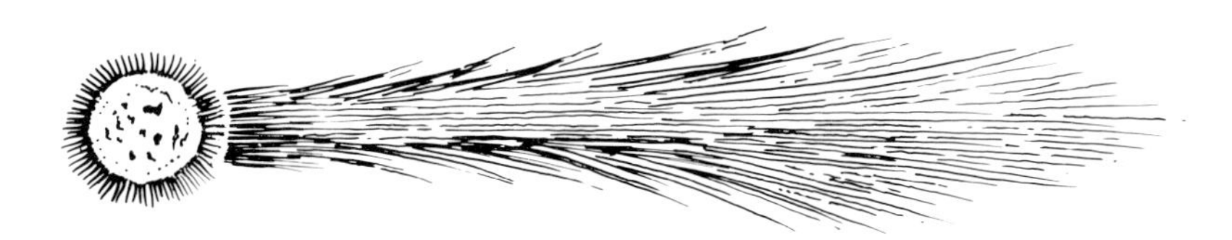

Halley's Comet in 1607: from an engraving in Johannes Hevelius, *Cometographia* (1688)

October 1 to 26. In spite of his work on the elliptical orbits of planets, Kepler apparently believed comets moved along straight lines. The comet was also recorded by Sir William Lower in South Wales, and very accurate positional measurements were taken by Thomas Harriot at Sion House near London from September 21 to October 23. The Chinese too had seen the comet already on September 21.

AD 1682

Date of perihelion: September 15
Least distance from Earth: 63 million km on September 1
Halley himself observed the comet from August 26 to September 10, but for his calculations of the orbits of comets and his conclusion that this was the same comet as had been seen already in 1531 and 1607 he relied on the observations made by the Astronomer Royal, John Flamsteed, and his assistant at Greenwich. He began calculating the orbits of comets and formulating his ideas on the periodic return of this particular one in 1695, but it was not until 1705 that his conclusions were published in Latin in the *Philosophical Transactions* of the Royal Society under the title *Astronomicae Cometicae Synopsis.* An English translation, *Synopsis of the Astronomy of Comets*, was published later in the same year.

CHAPTER 5

Present day uses of historic observations of Halley's Comet

In the previous chapters we have discussed historic observations of Halley's Comet from earliest times (240 BC) to the beginning of the telescopic era. Special emphasis has been placed on the more ancient sightings down to 12 BC. We need now to consider the usefulness of these records at the present day. It could be argued that the early observations themselves have their own special appeal, regardless of any practical value. In addition it is remarkable that it is possible to trace the history of any astronomical object back over more than 2000 years. However ancient and medieval observations of Halley's Comet have in fact played an important part in the study of the orbital motion of the comet, from the time of Halley himself.

The main reason why Halley's Comet attracts so much attention is the fact that it is so unusual. Many comets much brighter than this have appeared over the past centuries. However as far as is known all of these paid only brief visits to the inner solar system and may not return for many thousands or even millions of years. There is also a significant group of so-called 'short period' comets, orbiting the sun in anything from a few years to a few centuries. However, Halley's apart, these are generally invisible to the unaided eye throughout their course. Halley's Comet has been recorded at every single return since 240 BC.

Edmond Halley's discovery that the comet which now bears his name revolved in a periodic orbit around the sun taking approximately seventy-five years was the result of analysing careful observations of a variety of comets made since the European Renaissance. At the time the great wealth of Far Eastern records was virtually unknown. He calculated the orbits of more than twenty comets seen in the previous two centuries and found that although most had nothing in common three – dated AD 1531, 1607 and 1682 – were so alike as to be more than coincidental. This led him to propose that each sighting related to a return of the same comet. The analysis of historic records of comets had thus begun in earnest. Halley rather courageously anticipated the return of this comet 'about the end of the year 1758 or the beginning of 1759', even though he knew that he would not live long enough to witness it. Sixteen years after Halley's death the comet was duly sighted on Christmas Day 1758.

After Halley's time progress in extending the past history of his comet was slow, mainly due to the lack of useful early records. The situation changed dramatically when the French orientalist Edouard Biot, a former railway worker, made an extensive translation of Chinese cometary records in 1843. Biot's work was quickly put to use by astronomers, notably the Englishman John Hind. Mainly as the result of Hind's labour almost every return of Halley's Comet back to 12 BC was correctly identified with historic observations. Hind's method of investigation was fairly simple. Starting with a recent accurate date for Halley's Comet, he stepped back seventy-five years and looked for an observation which might correspond to the calculated motion of Halley's Comet. When he had made an identification he repeated this procedure until he could no longer trace the comet further back in time. Hind made a few incorrect identifications, but overall he revolutionised the whole history of the comet. It is to Hind that we owe the recognition of the comet of AD 1066, portrayed on the Bayeux Tapestry, as one of the numerous returns of Halley's Comet.

A much more analytical approach than that of Hind was developed by Philip Cowell and Andrew Crommelin not long before the last return of Halley's Comet in 1910. They used the technique known as numerical integration. Starting with recent observations of the comet, they calculated the whole orbit backwards in time, continually allowing for the perturbations of the comet's motion produced by the various planets, especially Jupiter and Saturn. Cowell and Crommelin considerably refined Hind's work and corrected several errors of identification. They even went so far as to suggest that the very brief Chinese record of 240 BC (discussed in chapter 3 above) was of Halley's Comet. A check on the agreement between their calculated dates for perihelion and those deduced from the observational records shows that back to AD 760 they were never more than about twenty days in error, while even as far back as 87 BC discrepancies did not exceed about fifty days.

There can be no doubt that Chinese observations have played a major part in the study of the past orbit of Halley's Comet. From 12 BC to AD 1378 all other observations tells us little that cannot be ascertained from the Chinese data alone. However the situation changed dramatically in AD 1456. In that year the Italian astronomer Paolo Toscanelli carefully measured the position of Halley's Comet to a fraction of a degree on every clear night for more than a month. He thus achieved far higher accuracy than the oriental observers who were still following traditional methods of observation. From this time on Far Eastern records are of only general interest, having been superseded by Western observers equipped with accurate measuring instruments and later on with telescopes. Nevertheless a great debt is due to the imperial astronomers of the Orient for keeping such a detailed and accurate record at a time when there was little astronomical activity in most other parts of the world. Without them the history of Halley's Comet would still be very incomplete.

Despite great advances in orbital theory and powerful methods of numerical integration using high speed computers, historic observations of Halley's Comet retain their importance today. Surprisingly enough some of the early observations are more accurate than calculation can achieve. The most interesting case relates to AD 837. In that year the comet made a remarkably close approach to the Earth, only about five million kilometres away. As a result the Earth considerably upset the motion of the comet and the modern astronomer, working backwards in time, cannot accurately calculate the dates of perihelion at previous returns without making allowance for observations made at this and other more ancient returns. It can be shown that had Halley's Comet in AD 837 arrived only four days later it would have almost reached inside the Moon's orbit, with a consequent huge alteration in its motion. This of course never occurred, but because we cannot calculate exactly what did happen in that year it is not possible to deduce the comet's previous motion on the basis of numerical integration alone. Fortunately Chinese observations made in AD 837 allow the date of perihelion to be deduced to within about 0.1 day. Modern investigators, in particular Donald K. Yeomans and Tao Kiang have made use of these and earlier observations to calculate the orbital motion to beyond 1000 BC. However, beyond about 87 BC the discrepancy between different theories becomes steadily worse. In fact by 500 BC there is disagreement over even the year in which the comet should have appeared. It is in dealing with problems such as these that the early observations prove invaluable. The recent discovery of Babylonian observations of Halley's Comet in 164 and 87 BC should contribute much to our understanding of the past orbit of this most famous of all comets.

Halley's Comet and Venus photographed from the Lowell Observatory, Arizona on 13 May 1910. Reproduced by permission of the Royal Astronomical Society, London

Further Reading

A. AABOE, 'Observation and theory in Babylonian astronomy', *Centaurus* 24 (1980) 14–35

D. H. CLARK and F. R. STEPHENSON, *The historical supernovae* (Oxford 1977)

H. H. DUBS, *History of the Former Han Dynasty* (3 vols, Baltimore 1938, 1944, 1955)

RUTH S. FREITAG, *Halley's Comet: A Bibliography* (Washington 1984)

EDMUND HALLEY, *A Synopsis of the Astronomy of Comets* (London 1705)

HO PENG YOKE, *The astronomical chapters of the Chin Shu* (Paris 1966)

P. HUBER, 'Ueber den Nullpunkt der babylonischen Ekliptik', *Centaurus* 5 (1958) 192–208

M. LOEWE, 'The Han view of comets', *Bulletin of the Museum of Far Eastern antiquities*, Stockholm, no. 52 (1980) 1–31

N. J. T. M. NEEDHAM, *Science and Civilization in China, 3: Mathematics and the Sciences of the Heavens and the Earth* (Cambridge 1959)

R. A. PARKER and W. H. DUBBERSTEIN, *Babylonian Chronology, 626 BC–AD 75* (Brown University Studies 19: Providence 1956)

T. G. PINCHES, J. N. STRASSMAIER, A. J. SACHS, and J. N. SCHAUMBERGER, *Late Babylonian astronomical and related texts* (Providence 1955)

E. REINER and D. PINGREE, *Babylonian Planetary Omens* (Bibliotheca Mesopotamica 2/2: Malibu 1981)

A. J. SACHS, 'A classification of the Babylonian astronomical tablets of the Seleucid period', *Journal of Cuneiform Studies* 2 (1948) 271–290

A. J. SACHS, 'Babylonian observational astronomy', in F. R. HODSON (ed.), *The place of astronomy in the ancient world* (Philosophical Transactions of the Royal Society A.276: London 1974) 43–50

E. H. SCHAFER, *Pacing the void: T'ang approaches to the stars* (Berkeley 1977)

F. R. STEPHENSON and D. H. CLARK, *Applications of Early Astronomical Records* (Bristol 1978)

F. R. STEPHENSON and K. K. C. YAU, 'Far Eastern observations of Halley's comet: 240 BC to AD 1368', *Journal of the British Interplanetary Society*, vol. 38 no. 5 (May 1985) 195–216

F. R. STEPHENSON, K. K. C. YAU and H. HUNGER, 'Records of Halley's comet on Babylonian tablets', *Nature*, vol. 314 no. 6012 (18 April 1985) 587–592

B. L. VAN DER WAERDEN, *Science awakening, II: the birth of astronomy* (Leiden and New York 1974)

D. K. YEOMANS and T. KIANG, 'The long-term motion of Comet Halley', *Monthly Notices of the Royal Astronomical Society* 197 (1981) 633–646